Cosmic Creation Story

Universal Creation, the First Beings, and Humanity's Remembered Gods

Janelle Staley

Cosmic Creation Story

First edition, 2026

ISBN: 979-8-234-19311-7

Reader's Note

This book presents my analysis of cosmic creation, the Lyran lineage, later genetically modified species, and the historical records that may preserve knowledge of these beings and events. I draw from astronomy, plasma physics, biology, genetics, archaeology, ancient writings, religious traditions, and surviving works of art. Scientific and historical information is presented alongside my interpretation of how these subjects connect within the larger creation story.

The lineage and off-world events are my personal cosmological interpretation, not established scientific or archaeological history. I distinguish those ideas from the findings of biology and astronomy and from what surviving texts and images actually record. Ancient descriptions are summarized unless quotation marks identify a translated passage; modern terms such as genetic engineering are my interpretation, not words translated from the tablets.

The biblical discussion is limited to pre-Christian Hebrew writings and related Second Temple material. The New Testament and later Christian doctrines are not used. Older textual variations, plural divine language, divine councils, and differences among translations are discussed within their historical contexts.

Bracketed numbers throughout the chapters refer to the complete source entries collected at the end of the book.

The photographs reproduce archaeological objects and sites; the cosmic scenes are interpretive illustrations. Their captions identify that difference. The source list at the end brings together the references used throughout the chapters.

Contents

Prologue - Reading the Oldest Memory

Humanity has developed different explanations for its beginning. Modern science describes a long physical and biological process through which matter formed stars, planets, living organisms, and eventually conscious beings. Ancient traditions describe humanity as the intentional creation of powerful beings who existed before human civilization. These accounts are usually treated as opposites. In my analysis, they may describe different stages of the same history.

Biological evolution explains how life adapts, diversifies, and becomes increasingly complex over time. It does not prevent an advanced civilization from later entering that process and modifying an organism that already existed. Matter, consciousness, evolution, and genetic intervention can therefore belong to one continuous creation story.

The oldest records were not written as modern scientific reports. They survive as temple texts, royal inscriptions, ritual recitations, family traditions, victory narratives, teaching stories, and poems copied by generations of scribes. Their language changed as governments, religions, and cultures changed. Later interpreters inherited fragments without always retaining the original circumstances that gave them meaning.

Alteration is not the same as invention. A memory can become surrounded by theology while still preserving an actual person, encounter, conflict, or technology that later societies no longer understand.

Imagine an isolated community witnessing an aircraft without possessing words for engines, alloys, electrical systems, or controlled pressure. Its observers would describe what the object did: it descended in brightness, roared like a storm, moved against the wind, opened, and released figures wearing unfamiliar coverings.

Later generations might remember it as a shining chariot, a thunderous bird, a moving throne, or a cloud containing fire. If the visitors selected rulers, taught specialized knowledge, interfered with reproduction, or imposed laws, the encounter would eventually become part of ancestry, ritual, government, and religion. Their instruments could become sacred emblems, while operating instructions could survive as ceremonial actions after their practical meaning was lost.

This possibility shaped the way I began connecting the oldest traditions. I noticed that ancient cultures repeatedly described differentiated celestial beings who possessed families, ranks, rivalries, vehicles, tools, territories, and specialized knowledge. They created or modified populations, transferred restricted knowledge, fought over authority, produced mixed descendants, imposed punishments, and selected survivors during catastrophes.

I interpret some of these accounts as historical memories preserved through the language available to the people who witnessed or inherited them. This does not require every deity to represent an extraterrestrial being or every unusual image to depict technology. It means that religious meaning, natural symbolism, and historical memory may exist together within the same tradition.

The overlapping inscriptions at Abydos show how a real ancient surface can accidentally resemble modern machinery. Elsewhere, complete narratives describe passengers, craft, reproductive processes, political disagreements, and the transmission of knowledge. These different kinds of record led me toward a cosmology that brings ancient memory into conversation with the physical and biological world.

The proposition

My reconstruction begins with consciousness rather than with a finished material universe. Source consciousness expressed itself through fields, radiance, electrically active plasma, vibration, rhythm, and geometry. These interactions produced increasingly stable cosmic structures, matter, planets, biological life, brains, and embodied awareness.

My account moves from that beginning into the Lyran civilization, its descendants, and the later human story. The connection I propose is not that every ancient god was the same being, or that every culture recorded the same encounter. It is that some traditions may preserve different parts of a larger history, reshaped by the communities that carried them.

Figure 1. An interpretive illustration of Source, radiance, plasma, and cosmic structure-not a telescope image or a demonstrated sequence of consciousness formation.

The lineage names

Lyrans, Aerians, Fraedians, Sateians, Faceasians, Ucefacians, and Heagface are names from the account I received, not names recovered from ancient inscriptions. My association of Faceasians with the Mesopotamian ruling gods is an interpretation, not a translation of Anunna or Anunnaki. I use Pleiadian for the guardian population's later adopted stellar region. These distinctions let the received names and the historical names remain visible without treating them as interchangeable.

PART I - SOURCE, PLASMA, AND LIVING FORM

CHAPTER 1

Source Before Substance

Before matter formed into stars, planets, or living bodies...

Before creation can be described, language has already divided what the theory holds to be whole. The moment we say Source, we imagine something. The moment we say before, we place it in time. The moment we say consciousness, we risk picturing a mind like our own, only larger. Source in my interpretation is not an elderly ruler seated beyond the universe. It is the undivided capacity for awareness, relation, and expression from which both observer and observed become possible. There is no external material waiting for Source to shape, because outside has not yet acquired meaning. Creation is differentiation within an original unity.

The first distinction is not a finished particle. It is polarity: the possibility that one region or condition can differ from another. Difference permits tension, and tension permits movement. Movement repeated becomes rhythm. Rhythm establishes interval, and interval makes proportion possible. Once proportion exists, pattern can return to itself. Circles, waves, spirals, standing modes, branching networks, and nested boundaries are not decorative additions placed upon reality. They are ways in which activity becomes coherent enough to persist.

Ancient traditions often place word, breath, tone, or light at the beginning. A modern reader can dismiss this as primitive metaphor, while an enthusiastic reader can make the opposite mistake and claim that an audible voice literally traveled through empty space. Both readings miss the deeper correspondence. A word is structured difference carried through a medium. Light is structured electromagnetic change. A musical tone is periodic motion. A spoken command also contains intention: pattern directed toward consequence. When early teachers joined creation with speech, they may have been recognizing that form begins when undivided capacity acquires ordered expression.

Physics does not presently identify intention at the origin of the universe. It describes relations among measurable quantities and tests models against observation. Cosmology reconstructs an early hot, dense condition and the subsequent expansion, cooling, formation of nuclei, atoms, stars, galaxies, and larger structures. Quantum field theory treats particles as excitations of underlying fields. These achievements do not prove that consciousness is fundamental, but they do unsettle the everyday picture of reality as inert pieces of substance. At the smallest described scales, the durable object gives way to probability, interaction, symmetry, and field behavior.

My reconstruction enters at that philosophical opening. It proposes that awareness is not an accidental light switched on by sufficiently complicated matter. Awareness is present as potential within Source, while individual minds are localized forms through which that potential becomes perspective. The universe does not awaken for the first time in the human skull; it becomes able to experience itself through increasingly integrated structures. A brain is crucial to human consciousness without necessarily being the ultimate origin of consciousness, just as a receiver can shape and localize a signal without creating every condition that makes transmission possible.

The hollow that makes movement possible

When I asked what the secret of the universe was, the answer I received was hollow. At first, hollow can sound like absence, as though the deepest truth were that nothing is there. My analysis moved in the opposite direction. Everything is hollow because openness is what makes movement possible. If there were no interval, no interior, no distance, and no receptive space, energy could not travel, vibration could not repeat, light could not cross a region, and one condition could not communicate with another. The hollow is not the failure of creation. It is the room creation requires.

I did not learn this interpretation in school. It came first through meditation, and I investigated the physical world afterward to see whether the insight had a real correspondence. I do not assign a fixed percentage to the emptiness of the universe, because science has no single measurement of how much of reality is 'nothing.' The answer changes according to scale and definition. On the largest scales, galaxies, gas, and dark matter gather into sheets and filaments around enormous cosmic voids that occupy most of the observable volume. Yet a cosmic void is underdense, not absolute nothingness. It can still contain thin gas, radiation, neutrinos, gravitational structure, dark matter, and quantum fields. In quantum field theory, fundamental fields extend throughout the universe, including regions ordinarily called vacuum.[26][28] What astronomy verifies is therefore not a universe made almost entirely of literal nothing, but a universe in which concentrated visible matter is separated by immense, extremely low-density spatial regions.

What appears solid at human scale is likewise not an absolutely packed substance without interior relation. Matter is organized through atoms, fields, charges, bonds, and probabilities. Atomic dimensions contain enormous separation relative to the nucleus, while electron distributions and quantum fields prevent that openness from behaving like an ordinary vacant room.[27] A table feels solid because electromagnetic and quantum interactions resist penetration, not because every possible location inside it is filled with an indivisible material block. Scientific language and my metaphysical language are not identical, but they touch at an important point: stable form does not eliminate space. Form depends upon ordered relations across it.

Vibration requires the possibility of change across an interval. A mechanical wave needs a medium whose parts can move and influence neighboring parts. Electromagnetic radiation does not require air, yet it still propagates through spacetime as changing electric and magnetic fields. Resonance requires boundaries, distances, timing, and repeated return. Even a standing wave is defined by the locations where motion is reinforced and the nodes where it is reduced. In this sense, space is active in every pattern. It permits difference, and difference permits movement.

This is why I do not interpret the void as dead nothingness. I interpret hollowness as receptive possibility. Source can differentiate because there is room for relation: here and there, movement and rest, pulse and interval, expansion and return. Polarity produces flow; flow produces rhythm; rhythm produces an interval; repeated interval produces proportion; and proportion becomes geometry. The universe can build because it is not sealed into one motionless, completely filled state.

The same principle continues into living systems. A cell needs an interior and exterior. Lungs need open chambers. Blood needs vessels. Sound depends upon cavities. The brain depends upon spaces, gradients, membranes, and timed gaps between signals. Communication exists because a signal can cross from one state or location to another. In my analysis, consciousness also needs this receptive quality: an opening in which information can arrive, be distinguished, and become experience.

The secret, then, is not simply that empty space exists. It is that hollowness makes connection, resonance, and transformation possible. The universe is able to move because it contains openness; it is able to form because movement can become patterned; and it is able to become conscious because patterned systems can receive, retain, compare, and respond. Space is not outside creation. Space is one of creation's essential conditions.

Sound, light, and space

Sound is mechanical vibration in a medium; it cannot travel through a vacuum as audible pressure. Light and radio waves are electromagnetic radiation and can cross a vacuum. When I connect creation with vibration, I mean patterned change, not an audible voice traveling through empty space. Ancient traditions of creative speech and illumination resonate with this interpretation, while the physical mechanisms of sound and light remain distinct.

CHAPTER 2

Plasma: The First Responsive Architecture

Charged matter, magnetic fields, and the emergence of structure

If Source is undivided capacity and the first creative movement is patterned differentiation, plasma becomes important because it is matter behaving relationally in an unusually visible way. A neutral gas can often be approximated as particles moving and colliding. In plasma, charged particles participate in collective electrical and magnetic behavior. A disturbance can organize motion far beyond a single collision. Currents form magnetic fields; magnetic fields redirect currents. Boundaries arise between regions with different electrical properties. Filaments twist through luminous volumes. Waves carry energy and information. Sudden reconnection can release stored magnetic energy with spectacular force.

This is not the exotic exception it may appear to be from the surface of Earth. Stars are plasma. Stellar winds are plasma. Planetary magnetospheres interact with plasma. Auroras reveal charged particles guided by magnetic geometry. Hot gas between galaxies and inside clusters can be ionized. Much of what astronomy sees shining is matter in a plasma state, even though dark matter, neutral gas, dust, compact objects, and the geometry of spacetime are also essential to a complete cosmic account. Calling the universe plasma-rich is sound. Claiming that plasma alone replaces gravity, relativity, or standard cosmology is not required by my analysis.

The attraction of plasma for a consciousness-centered cosmology comes from its responsiveness. A field-responsive medium can create form without first becoming solid. It can establish a boundary, maintain a current channel, divide into cells, and reorganize abruptly when conditions cross a threshold. The resulting structures are not alive in the biological sense. They do not possess demonstrated intention, metabolism, reproduction, or heritable evolution. Nevertheless, they offer an important bridge:

organized form does not begin only when rock hardens or DNA appears. Dynamic matter can hold structure temporarily, transmit disturbances, and couple widely separated regions.

In my reconstruction, Source first expresses through fields and plasma because these provide an active canvas. Light makes energetic transitions visible. Electromagnetic interaction creates attraction, repulsion, rotation, confinement, and release. Resonant modes favor some patterns over others. Where a configuration channels energy efficiently, it may persist; where it cannot, it dissolves. This is the earliest meaning of creation as selection, not a conscious artisan choosing a finished animal, but ordered potential exploring forms and retaining coherence.

The step from responsive plasma to durable matter requires cooling and layered physical processes. In the early universe, electrons and nuclei could not remain joined while radiation was sufficiently energetic. As expansion lowered the temperature, neutral atoms formed and light traveled more freely. Gravity gathered matter into denser regions. Stars ignited nuclear fusion, transforming light elements and later distributing heavier ones through winds, eruptions, collisions, and stellar death. Carbon, oxygen, silicon, iron, phosphorus, and the other materials of planets and bodies emerged through this history. Plasma is therefore not a substitute for chemistry. It is the first great responsive architecture within a sequence that eventually makes chemistry possible.

Currents and filaments

Plasma contains free electrons and ions whose collective behavior supports waves, sheaths, instabilities, and current channels. NASA's observations of the Sun and solar wind reveal these interactions on immense scales.[1] Laboratory plasma research likewise examines self-organized electromagnetic structures.[2] Birkeland currents flow along magnetic-field lines; twisted or pinched forms depend on the surrounding conditions. Not every cosmic spiral is a Birkeland current, and the illustration below is a conceptual scene rather than a record of one measured configuration.

Figure 2. Conceptual plasma artwork showing luminous filaments and magnetic organization. The shapes illustrate the theme of patterned currents; they are not a measured model of the universe's formation.

Consciousness and responsiveness

Where does consciousness enter? The cautious scientific answer is that no experiment demonstrates awareness in cosmic plasma. The reconstructive answer is subtler: awareness need not belong to each glowing filament as a little personality. Plasma may have been the earliest material through which fundamental consciousness could differentiate, communicate, and sustain patterned activity. Individual consciousness would appear much later as more complex systems created bounded perspectives. Under this view, plasma is not a cosmic brain. It is the first field-responsive fabric capable of carrying the movement from undivided potential toward articulated form.

CHAPTER 3

Geometry, Resonance, and Living Form

From physical pattern to organisms, nervous systems, and awareness

Creation becomes visible as geometry because every enduring system must answer the same questions: Where is its boundary? How does energy enter and leave? How are stresses distributed? How does one part remain coordinated with another? A soap film, a spiral galaxy, a shell, a branching river, a vascular network, and a magnetic filament do not share one simple cause. They do show that form is constrained. Matter cannot arrange itself in every imaginable way with equal stability. Repeated shapes emerge because certain relations solve recurring physical problems.

Sacred geometry is often presented as though recognizing a spiral or proportional relationship proves a hidden designer. The evidence does not permit that shortcut. Spirals can emerge from rotation and growth; hexagonal packing can minimize space; branching networks can distribute flow; bilateral symmetry can serve directed movement. Natural processes are capable of producing beauty without a craftsman placing each line. Yet lawful self-organization is itself part of the creation story. Source need not interrupt nature to be present within it if nature is the unfolding grammar of Source.

Resonance adds time to geometry. A standing wave creates stable nodes because repeated motion reinforces some positions and cancels others. A resonant cavity selects particular modes. An atom absorbs and emits specific energies. Molecules vibrate in characteristic ways. Biological systems also rely upon rhythm: gene expression cycles, neural oscillations, heartbeats, breath, hormonal pulses, sleep-wake timing, and coordinated movement. The word resonance is easily abused in spiritual language, but its disciplined meaning remains powerful. Systems respond selectively.

Timing can unite small forces into large effects. Coherence changes what a collection of parts can do.

Life deepens this principle by storing successful organization. A living cell maintains an interior different from its surroundings. Its membrane is not a wall of total separation; it is an intelligent boundary in the biological sense, regulating exchange through evolved molecular machinery. DNA and related cellular systems preserve instructions, but genes do not act alone. Development depends on chemical gradients, electrical states, mechanical forces, timing, tissue interaction, and environmental conditions. A body is not assembled like a statue from a single blueprint. It grows through communication among parts.

Information becomes inheritance

A reproducing system carries information into later generations. On Earth, mutation, natural selection, genetic drift, gene flow, and symbiosis help explain the diversity of life. No single geometric shape specifies an organism. Molecular structure, cellular chemistry, development, and environmental conditions interact to produce bodies that can maintain themselves and reproduce.

From cells to nervous systems

Life requires boundaries. A cell could not exist without a membrane separating its internal chemistry from the surrounding environment. That boundary was not complete isolation; it selectively allowed energy, nutrients, waste, and information to pass through. The earliest organisms survived by detecting differences around light and darkness, warmth and cold, nutrients and toxins, safety and danger. Even before brains existed, living cells could sense conditions, respond to chemical signals, preserve information, and adjust their behavior.

Over generations, organisms changed through mutation, inheritance, natural selection, genetic drift, and environmental pressure. Traits that helped organisms survive and reproduce were more likely to continue. Different environments encouraged different biological solutions. Water favored bodies capable of controlling buoyancy and detecting pressure. Land required support against gravity, protection from dehydration, and new forms of movement. Darkness favored sensitivity to vibration, heat, scent, or faint light. Predation encouraged speed, camouflage, defensive structures, cooperation, memory, and anticipation.

Evolution was not a straight ladder moving inevitably toward human intelligence. It formed a branching tree in which different organisms developed different kinds of awareness suited to their environments. An octopus, bird, dolphin, primate, insect colony, and human do not experience or process the world in the same way. Intelligence can appear through tool use, communication, navigation, memory, cooperation, problem-solving, emotional recognition, or the ability to predict another being's behavior.

As organisms became more mobile and their environments became more complicated, faster coordination became advantageous. Networks of specialized cells developed to transmit electrical and chemical signals throughout the body. Simple nerve nets could coordinate movement and sensation. More centralized nervous systems could compare information from different senses, store previous experiences, select between possible actions, and organize increasingly complex behavior.

The brain emerged gradually from this biological need to connect perception with action. It did not appear suddenly as a finished organ. Earlier nervous systems already performed parts of the process later concentrated within brains. Over time, sensory information, memory, emotion, body regulation, movement, and decision-making became increasingly integrated. An organism could do more than react to the immediate moment; it could remember what happened, anticipate what might happen next, and alter its behavior accordingly.

The brain and conscious experience

In my interpretation, this increasing integration created a biological opening through which consciousness could become localized. I do not interpret the brain as necessarily creating consciousness from nothing. I view it as a living receiver, organizer, and translator that allows a broader field of consciousness to become an individual perspective. The structure and condition of the brain influence what can be perceived, remembered, communicated, and expressed, just as the design of a receiver determines which signals it can detect and how clearly it can reproduce them.

This does not mean that brain activity and consciousness are unrelated. Damage to the brain can alter memory, perception, personality, language, and awareness. Sleep, medication, injury, disease, electrical activity, and changes in brain chemistry can all affect conscious experience. These facts demonstrate that the brain is essential to the human form of consciousness. They do not, by themselves, settle the deeper philosophical question of whether consciousness originates entirely inside the brain or whether the brain organizes something more fundamental.

Biological consciousness may have developed by degrees. Basic organisms detected and responded to their surroundings. More complex animals combined sensation with memory. Social organisms needed to distinguish between themselves and others, interpret intentions, recognize relationships, and predict behavior. Increasingly integrated brains could construct an internal model of the body, the environment, and eventually the self experiencing both.

At some point, awareness became capable of reflecting upon itself. A conscious being could not only feel pain, seek food, avoid danger, or recognize another individual; it could ask what it was, where it came from, and how reality worked. The universe had formed matter, matter had organized into life, and life had developed structures through which the universe could consciously observe and examine itself.

In my interpretation, the Lyrans belong after this biological development, not outside it. They arose within a planetary biosphere and became capable of studying the processes that had shaped them. The transition from an organism responding to its surroundings to a civilization deliberately changing living inheritance is the threshold at which their story begins.

PART II - THE LYRAN CIVILIZATION AND ITS DESCENDANTS

CHAPTER 4

The Lyrans: The Root Civilization

Perception, technology, and the decision to preserve life beyond one world

In my analysis, the Lyrans were the first advanced biological civilization to develop from the original union of Source consciousness, plasma, matter, and living form. They represent the root population from which the later engineered species descended. They were not feline beings, as they are frequently portrayed in modern online accounts. They were humanoid in appearance, generally taller and more physically substantial than present-day humans, with darker or tan skin adapted to the conditions of their planet.

Their place among the stars

Lyra marks a direction in Earth's sky, not one gravitationally bound star system. Its stars occupy different distances and have different ages. Vega is therefore not the sun of the civilization in my account. The system I have in mind is older and unidentified. Kepler-444, in the direction of Lyra, provides an astronomical comparison: its small planets formed around a star estimated to be roughly eleven billion years old.[29] That discovery establishes the antiquity possible for planetary systems, not the location of a lost Lyran world. No known debris field has been identified as the remains of the planet described here.

Biology and Source perception

Lyran biology retained a much stronger connection to the fields and energies from which physical reality emerged. Their senses extended beyond ordinary sight, hearing, touch, taste, and smell. They could perceive subtle electrical, electromagnetic, vibrational, and microscopic interactions that later species could detect only partially or measure through instruments. Their awareness of Source was therefore not limited to a religious belief. It was an experienced part of their physical reality.

This wider perception influenced the development of Lyran intelligence. They understood that matter was not motionless or isolated, but organized through energy, charge, vibration, geometry, and fields. Because they could perceive more of these interactions directly, they learned how to manipulate physical laws with greater precision than modern humanity. Their science, consciousness, and technology developed together rather than being treated as separate areas of knowledge.

The Lyrans created Source-based energy systems by learning how to direct and organize energies already present within the universe. They did not create Source itself. They created technologies that accessed, concentrated, converted, and applied its available expressions. Their spacecraft, biological laboratories, communication systems, and methods of planetary exploration were extensions of this understanding.

The reason for the experiments

Like humans, the Lyrans understood that no planet remains permanently safe. Meteor impacts, changes in stars, geological instability, atmospheric loss, disease, resource depletion, and conflict can threaten even an advanced civilization. They developed space travel and planetary-seeding programs to protect the continuation of their population and knowledge.

The initial biological material came from the Lyrans' own world. Like Earth, that planet supported many forms of life. Reptilian, avian, aquatic, rodent, batlike, and cave-insect organisms carried different adaptations to its environments. In my interpretation, the Lyrans combined selected traits with their own inheritance to explore survival beyond their large humanoid form, develop specialized workers and guardians, and test improvements to their own capacities.

The experiments involved whole organisms: metabolism, development, reproduction, immunity, and nervous systems had to remain compatible. A useful sense or protective feature could come with new limitations. Some lines became independent peoples; others remained restricted experiments. Their common origin did not give them identical intelligence, needs, or destinies.

Figure 3. My imagined Lyran laboratory, using biological material from their home planet. This is interpretive artwork, not an archaeological reconstruction based on recovered equipment.

Spaceflight and preparation

The Lyrans were an advanced technological civilization as well as a spiritually developed one. Their capabilities extended far beyond present human spaceflight. They could travel between worlds, transport living populations and biological material, survey planetary environments, and plan settlements across astronomical distances. Their genetic knowledge belonged inside this larger technical system; creating a viable lineage for another planet required transportation, habitat analysis, controlled reproduction, ecological preparation, and continuity across generations.

Preparation preceded the war. Surveying other worlds and establishing routes created options if the home planet became uninhabitable. Genetic adaptation and transport belonged to the same effort: a population needed both a body able to survive and a destination that could sustain it. This survival motive existed alongside the unequal relationship between designers and the beings whose purposes they selected.

Lyran society was technologically advanced because its people understood the relationship between consciousness and physical law. Their Source connection gave them broader perception, but it did not make every decision infallible. Their genetic programs reveal both foresight and risk. They attempted to preserve intelligent life across different environments, yet they also created populations whose development could exceed their control.

The following portraits describe the distinct branches once. Their possible appearances in Earth traditions are discussed with the relevant cultures later in the book. Only the Faceasian branch is connected in my interpretation with the genetic project that produced humanity; resemblance to another lineage does not imply participation in that project.

CHAPTER 5

The First Modified Lineages

Reptilian, avian, aquatic, and rodent adaptations

Reptilians: The First Modified Lineage

Survival ability, inherited technology, and the growth of independent power

In my analysis, the Reptilians were the first major intelligent population developed by combining Lyran inheritance with reptilian life from the Lyran home planet. The experiment sought resilient workers and tested whether protective and survival traits could extend the capacities of the original population.

Body and perception

Reptilian biology was particularly effective in demanding environments. Protective skin and efficient metabolism reduced vulnerability to exposure and limited resources. Chemical and thermal sensitivity helped them detect changes in their surroundings, while rapid defensive responses allowed them to react before slower organisms could evaluate a threat consciously.

Their Lyran inheritance gave them advanced reasoning, memory, communication, and the ability to understand the scientific knowledge taught by their creators. However, the increasing influence of specialized planetary genetics reduced portions of the broader Source perception retained by the original Lyrans. Reptilian intelligence consequently developed with a stronger emphasis on immediate conditions, physical survival, territorial security, strategy, and control.

These traits did not automatically make every Reptilian violent or destructive. They created a biological and psychological foundation that could support patience, protection, endurance, and disciplined planning. Under leadership centered on fear and expansion, however, those same abilities could be redirected toward surveillance, domination, conquest, and the elimination of perceived threats.

An independent power

As the Reptilian population expanded, it no longer accepted permanent Lyran authority. The beings the Lyrans had created learned their technology, scientific principles, transportation systems, defensive methods, and organizational structures. They eventually developed their own leadership and began using inherited Lyran knowledge to pursue independent territorial power.

The conflict was partly political and partly ideological. The Lyrans viewed planetary expansion as a way to preserve life, knowledge, and conscious development against natural catastrophes. Reptilian ruling factions increasingly treated survival as a struggle in which security required control over territory, resources, technology, and other populations. Freedom appeared unpredictable; domination appeared protective.

The creators could no longer control the civilization that had developed from their experiment. The Reptilians understood enough Lyran science to challenge them directly, and their survival-oriented biology made them highly effective in prolonged conflict. What began as a created population seeking independence developed into a military power capable of attacking its own ancestral civilization.

The later name "Draconian"

I use Reptilian as the name within this account. Draconian appears in later esoteric descriptions, sometimes linked to dragons and sometimes to the constellation Draco. I regard the dragonlike appearance as a possible reason for that later label, not as an established etymology or proof of a birthplace in Draco. The population's origin in my interpretation remains the Lyran experiment; its later conflict belongs to the shared history of the dispersal.

Aerians: The Avian Lineage

Distance, motion, and a three-dimensional experience of the world

In my interpretation, the Aerian line joined Lyran inheritance with avian biology already present on the Lyran planet. Its distinguishing development was a sensory and bodily organization suited to aerial movement, navigation, and observation. Flight involves coordinated changes in the skeleton, muscles, respiration, circulation, balance, and energy use, not simply the addition of wings.

Aerial perception

Aerian perception was organized around distance, motion, direction, and spatial relationships. Their visual systems allowed them to recognize movement and patterns across large areas, while their balance and navigation systems supported controlled movement through open space. Depending on the avian genetics used, some populations may also have perceived ultraviolet light, polarized light, magnetic orientation, or subtle changes in atmospheric pressure that remained outside ordinary Lyran perception.

Their aerial perspective affected more than physical movement. A being capable of observing landscapes from above would understand boundaries, settlements, migration routes, weather systems, approaching danger, and the relationship between distant locations differently from a ground-dwelling population. This gave the Aerians a natural advantage in navigation, exploration, mapping, observation, communication, and the coordination of activity across wide territories.

The Lyran contribution gave these sensory abilities a higher level of conscious interpretation. The Aerians did not merely react to currents and directional cues instinctively; they could study them, communicate what they perceived, and incorporate that knowledge into travel and technology. Their ability to combine direct environmental awareness with inherited Lyran science made them effective navigators and observers.

As with the other modified populations, specialization also produced tradeoffs. A body adapted for aerial movement required different energy use, skeletal support, environmental conditions, and methods of protection. Some parts of the original Lyran perceptual range may have narrowed as biological resources became concentrated on vision, balance, atmospheric sensing, navigation, and rapid spatial processing.

Life above the surface

Aerian communities developed around mobility and a three-dimensional understanding of territory. Settlements may have been positioned in elevated regions, cliffs, large structures, or other locations that allowed safe access to open air. Travel routes would not have been limited to the surface, and boundaries may have been understood through altitude and airspace as well as land.

Their societies likely valued observation, navigation, communication, and the ability to recognize changes before they reached the population below. Individuals with exceptional long-range perception could serve as scouts, messengers, guides, guardians, or teachers. Their inherited technology could further extend these natural capacities through instruments designed for mapping, communication, atmospheric travel, and the detection of distant activity.

The Aerians were not automatically peaceful or morally superior because they possessed a wider visual perspective. The same abilities that supported exploration and protection could also be used for surveillance, territorial advantage, or warfare. Their history would have been shaped by how different groups chose to use the specialized perception and mobility created through their biology.

The wider dispersal separated Aerian communities into different environments. Their later development need not have preserved one unchanging appearance or one political allegiance. I connect their history chiefly with mobility and perception, rather than with the direct ancestry of humans.

Fraedians: The Aquatic Lineage

Pressure, current, and perception within a surrounding medium

In my analysis, the Fraedians originated from Lyran inheritance combined with aquatic or fish-derived organisms from the waters of the Lyran home world. Their development opened environments in which an unchanged surface body would be limited. Breathing, circulation, movement, temperature control, and perception had to function together in water.

Sensing the surrounding water

Fraedian perception developed within a medium that continuously carries motion and information. Water transmits vibration, pressure, sound, chemicals, temperature, and electrical activity differently from air. A Fraedian could detect approaching organisms, environmental changes, distant disturbances, and the movement of groups without relying entirely on ordinary vision.

Aquatic animals on Earth demonstrate how varied such perception can become. Fish use lateral-line systems to detect changes in water movement and pressure. Sharks and rays detect weak electrical fields produced by other organisms. Many aquatic mammals communicate across long distances through sound, while other species sense chemical gradients, temperature boundaries, or changes in current. In my analysis, the Fraedians integrated comparable abilities with the conscious reasoning inherited from the Lyrans.

I associate the Fraedians with awareness of emotional currents. In biological terms, breathing, movement, posture, and chemical release can reveal an animal's condition; some aquatic species detect such cues. Extending that sensitivity into a deliberate perception of emotion is my interpretation, not an established ability to read thoughts through water.

Their specialized biology allowed them to interpret a world that surface beings would experience only partially. At the same time, adaptation to pressure, water chemistry, depth, and limited light could make Fraedians dependent on particular aquatic conditions. Their technology would have needed to regulate living environments, protect them during travel, and translate between aquatic and atmospheric settings.

Aquatic communities

Fraedian settlements developed within oceans, underwater caverns, protected basins, and other stable aquatic environments. Their architecture would have been shaped by pressure, current, buoyancy, visibility, and the movement of water rather than by the surface-based assumptions of later human construction. Structures could work with surrounding currents, using the natural movement of the environment for transportation, communication, energy transfer, or circulation.

Their societies likely placed strong value on coordination because changes within a shared body of water could be sensed throughout a community. Communication may have included sound, vibration, movement, electrical signaling, and other patterns transmitted through the surrounding medium. This could create a strong awareness of group conditions while still allowing individual thought and identity.

The same sensitivity that supported cooperation could also make privacy and separation more difficult. Individuals might be continuously aware of the movements and emotional states of those around them. Fraedian culture would therefore need ways of balancing collective awareness with personal boundaries.

Their inherited Lyran knowledge allowed them to develop technologies appropriate for underwater construction, navigation, environmental regulation, transportation, and communication. Some of these systems may have allowed them to operate beyond their native waters through enclosed craft or controlled fluid environments.

Some Fraedian populations may have traveled with Lyran survivors or other allied groups, while others remained within established oceanic civilizations. Their dependence on aquatic conditions would have shaped which planets they could inhabit and what transportation systems were required to move them safely. Over time, separated groups would have continued adapting to different oceans, producing further biological and cultural variation within the lineage.

Sateians: The Resilience Lineage

Small bodies, flexible behavior, and recovery after disruption

In my interpretation, the Sateians developed from a resilience experiment using Lyran inheritance and rodent stock from the home planet. Their purpose centered on rapid adaptation and population continuity under difficult conditions. Rodent biology offered examples of learning, flexible food use, concealment, and reproductive recovery.

Adaptability and its costs

Sateian biology emphasized alertness, speed, environmental awareness, and rapid learning. Their senses were adapted to detecting nearby movement, chemical traces, changes in available food, concealed threats, and safe passage through crowded or enclosed environments. Strong hearing and smell would have helped them function in darkness or in locations where vision was limited.

Their intelligence was practical and responsive. Rather than depending entirely on fixed plans, they could study immediate conditions, improvise with available materials, and change their behavior quickly. This allowed Sateian populations to survive where larger and more specialized beings might struggle.

Their rate of reproduction and development also made them valuable to a civilization concerned with population survival. A Sateian settlement could recover more quickly after disease, resource loss, natural disaster, or war. This made the lineage one of the earliest tests of whether intelligent life could be designed to withstand catastrophic population decline.

These advantages also created difficulties. Rapid reproduction could exceed available resources, especially when populations entered unfamiliar environments. Constant alertness could increase competition, suspicion, or fear. Their ability to survive within the margins of other societies could cause them to be treated as unwanted populations even when their behavior reflected adaptation rather than hostility.

Communities and continuity

Sateian communities developed around cooperation, shared information, concealed infrastructure, and the efficient use of limited resources. They could establish settlements quickly and maintain them without the large construction systems required by some other populations. Their ability to repurpose materials and adjust to changing conditions reduced their dependence on centralized supply networks.

Knowledge was likely transmitted through observation, imitation, and direct practical teaching. Younger individuals could learn survival patterns rapidly, allowing useful behavior to spread throughout a population within a short period. Their inherited Lyran intelligence transformed instinctive adaptability into deliberate planning, toolmaking, and social organization.

Sateian societies may have favored distributed communities rather than a single concentrated center. If one settlement failed, others could survive elsewhere and continue the population. This decentralized structure made them difficult to eliminate but could also make unified political organization more challenging.

Their independence could become a source of tension with the makers. A population able to find shelter, repurpose resources, and teach practical survival skills would be difficult to keep permanently dependent. I associate their survival through the later instability with that flexibility, without assigning them a role in the Earth genetic project.

Their survival did not depend upon maintaining the full technological structure of the original Lyran civilization. Even when advanced systems were lost, Sateian communities could preserve essential knowledge through practical use and social learning. This gave the lineage continuity during periods when more centralized civilizations collapsed.

CHAPTER 6

Specialized Lineages and the Limits of Adaptation

Engineering, guardianship, and experiments with uncertain futures

Faceasians: The Ruling and Engineering Lineage

Specialized knowledge, institutions, and concentrated authority

In my analysis, the Faceasians developed from a Lyran-derived population shaped for administration, engineering, scientific specialization, and the coordination of large societies. They inherited advanced knowledge from the earlier Lyran civilization, including space travel, energy systems, planetary science, biological engineering, construction, resource extraction, and the management of long-term settlements.

Unlike lineages created primarily for flight, aquatic survival, rapid population recovery, or subterranean life, the Faceasians were developed to organize complex civilizations. Their abilities centered on planning, technical memory, command structures, biological research, infrastructure, and the division of work among specialized groups.

Their connection to Source perception was weaker than that of the original Lyrans, but they retained enough inherited knowledge to understand and manipulate many of the physical forces their ancestors had perceived more directly. What the earliest Lyrans could both sense and study, the Faceasians increasingly understood through preserved science, technology, instruments, and written records.

This distinction shaped their civilization. Knowledge became concentrated within specialist groups and ruling families rather than remaining equally accessible to the entire population. Those who controlled spacecraft, laboratories, energy systems, genetic archives, and planetary resources also controlled the populations that depended on those systems.

Technical specialization

Faceasian bodies remained more humanoid than many of the other modified lineages. Their primary changes supported intelligence, memory, communication, longevity, planning, and the ability to coordinate large technical operations. Their brains were capable of holding complex systems of information and understanding how biology, energy, machinery, environment, and social organization interacted.

Their civilization depended on specialization. Geneticists, medical workers, engineers, pilots, builders, record keepers, resource specialists, administrators, and labor supervisors performed different functions within a ranked system. Large projects were not completed by a single ruler or scientist. They required teams operating within an established command structure.

In my interpretation, the name Enki later came to stand for the leading scientist within a much larger operation. Research, reproductive work, transport, medicine, and construction required teams. This distinction matters to the human story: the actions attributed to one named maker can also represent work undertaken under that maker's authority.

The Faceasians' intelligence did not guarantee wisdom or equal treatment of other conscious populations. Their ability to design biological workers encouraged some rulers to view living beings as resources created for assigned functions. Intelligence became separated from Source awareness, allowing technical achievement to continue even when empathy, shared consciousness, and respect for individual freedom had diminished.

Government and the ruling family

Faceasian society was organized through families, councils, ranks, territories, and assigned responsibilities. Authority was concentrated among powerful ruling lines whose members governed settlements, directed workers, controlled resources, and competed for influence.

I connect this lineage with the Mesopotamian family remembered through Anu, Enki, Enlil, and Marduk. In the arrangement I use, Anu is senior authority, Enki and Enlil are brothers with different responsibilities, and Marduk is a son of Enki/Ea. Mesopotamian genealogies vary across periods and cities; that family arrangement is not uniform in every ancient source. Enki's role in my account is scientific and protective, while Enlil's is administrative and coercive. Their literary counterparts and Marduk's rise in Babylon are developed in the Mesopotamian chapter.[12][15][16]

The Earth episode follows from this institutional history. The Faceasians possessed the knowledge to undertake a biological project, but its purpose was defined within a system of resources, work, hierarchy, and political disagreement. Technical competence did not settle the question of how a conscious worker deserved to be treated.

Pleiadians: The Guardian Lineage

Ultrasonic hearing, infrared sensitivity, and an adopted stellar home

In my analysis, the beings later called Pleiadians began as a guardian experiment using Lyran inheritance and batlike organisms from the Lyran home planet. The intended specialization was ultrasonic hearing and infrared sensitivity: protection through detecting activity that ordinary sight and hearing missed. Their later name identifies an adopted region, not the location of that experiment.

Hearing and thermal perception

Ultrasonic hearing allowed this lineage to detect sound frequencies above the normal human hearing range. Through the movement of sound and the timing of returning echoes, they could determine the distance, direction, size, position, and movement of surrounding objects even when those objects could not be seen.

This perception created an internal spatial awareness of the environment. Walls, passageways, approaching beings, moving equipment, and changes in open space could be detected through sound reflection. The guardians were therefore not dependent upon visible light to understand what surrounded them.

Their infrared sensitivity allowed them to detect differences in heat. Living organisms, operating machinery, energy systems, spacecraft components, and recently disturbed surfaces all produce or retain thermal patterns. This ability helped the guardians locate concealed beings, identify biological activity, monitor equipment, and detect abnormal changes before they became visible.

Earth biology supplies partial parallels. Many bats use ultrasonic echolocation, whereas vampire bats detect radiant heat through specialized facial structures involving a heat-sensitive form of TRPV1.[30] These are not abilities shared identically by all bats, and heat detection is not ordinary optical vision. I interpret the guardian line as combining and extending comparable functions with conscious reasoning. Heat or sound would still need a pathway to the receiver; neither ability implies seeing through every solid obstruction.

The Pleiadians did not experience these signals only as automatic instincts. Their Lyran-derived brains allowed them to interpret, compare, remember, and communicate what they perceived. They could distinguish ordinary environmental activity from a possible threat and coordinate their response with others.

Their technology extended these natural senses. Instruments could amplify weak signals, translate additional frequencies, monitor distant locations, record changes, and communicate warnings across settlements or spacecraft. Their biological perception and technology worked together as one protective system.

Guardians and the limits of protection

Their role was protective rather than administrative. They were not created to govern planetary populations or manage labor systems in the manner later associated with the Faceasians. Their primary function was awareness: detecting what others could not perceive and warning the larger population before danger arrived.

As a conscious species, however, they eventually became more than the purpose assigned by their creators. They developed their own families, communities, knowledge, technology, and understanding of their place within the wider lineage. A population created to protect others became capable of defining what protection meant for itself.

Society and Development

Pleiadian society developed around observation, preparedness, communication, and coordinated protection. Individuals were highly aware of movement, sound, heat, and subtle changes within their surroundings. This made concealment more difficult and encouraged forms of social communication based on precise sensory information.

Guardian groups could divide responsibility across different areas and frequencies. Some individuals might monitor distant or external activity, while others watched biological conditions, machinery, transportation systems, or environmental changes. Information could then be combined into a more complete understanding of the surrounding situation.

Their heightened senses also required regulation. Constant exposure to sound, heat, movement, and activity could overwhelm the nervous system without the ability to filter unnecessary information. Pleiadian development therefore likely included training in attention, sensory control, rest, and the conscious selection of relevant signals.

Although their purpose emphasized protection, their abilities could also be misused. The same senses that detected danger could support surveillance, pursuit, or control. The character of a guardian population depended upon who defined the threat and who controlled the information. Different Pleiadian communities could therefore develop different political loyalties and approaches to protection.

Settlement in the Pleiades

In my account, members of this population later settled in the Pleiades and became known by that region. The Pleiades is an open star cluster within Taurus, not a separate constellation or a single solar system. Its young stellar population does not establish the age of the ancestry I propose: migration into a younger region is different from biological origin there. The settlement itself belongs to my interpretation, not to an astronomical discovery of inhabited planets.

Ancient cultures carefully observed the Pleiades and connected the cluster with seasonal cycles, navigation, agriculture, ancestry, and groups of celestial persons. These traditions establish the importance of the star cluster but do not clearly identify the guardian species described here. The lineage's history comes from my interpretation of the wider Lyran record rather than from an ancient Earth text naming these beings as Pleiadians.

Ucefacians: A Smaller Form, a Local Future

Adaptation could preserve a body without protecting an entire world

In my analysis, the Ucefacians developed from genetic material left over from Lyran experiments in physical survival and planetary adaptation. The original Lyrans were tall, giantlike humanoids whose large bodies were suited to the conditions of their home planet. Their size became a limitation when they began exploring other worlds with different gravity, atmospheres, available resources, and living environments.

A larger body requires more food, water, oxygen, physical space, and energy. It also places greater demands on the heart, circulation, skeleton, muscles, and temperature-regulation systems. A planet with stronger gravity could place excessive strain on a large Lyran body, while a world with limited resources might be unable to support a substantial population of beings of that size.

The Lyrans therefore experimented with reducing their physical form while attempting to preserve intelligence, mobility, reproduction, memory, and enough of the original biological structure to maintain the continuation of their species. The objective was survival adaptation rather than the creation of a completely unrelated population.

Not every genetic variation was incorporated into the primary Lyran line. Some of the smaller-bodied forms developed separately from the larger population. This redirected experimental stock became the foundation of the beings later called Ucefacians.

Body and adaptation

The Ucefacians retained a humanoid Lyran foundation but possessed smaller and more resource-efficient bodies. Their reduced size allowed them to survive with less food, water, oxygen, and living space. It also made physical movement easier under gravitational conditions that would have placed greater stress on the much larger original Lyrans.

Their metabolism, circulation, respiration, skeletal structure, and muscle distribution had to be adjusted together. Simply reducing the dimensions of the body would not have produced a stable organism. Every organ and biological system needed to support the altered scale while preserving brain function, reproduction, immunity, movement, and long-term health.

The Ucefacians retained Lyran intelligence and portions of inherited technical knowledge. However, the genetic changes that reduced and stabilized their bodies also altered other inherited capacities. As occurred with many of the modified lineages, some aspects of the original Lyran Source perception became weaker, inactive, or more narrowly expressed.

A nearby home

The Ucefacians did not disperse across distant regions of the galaxy. The population moved only to a nearby planet within the area humans now identify as the Lyra constellation. This planet became their only established home beyond the location where the genetic experimentation had begun.

The location is in the direction now called Lyra, but the star and planet have not been identified. The Ucefacians remained a local branch rather than joining the distant migrations. Their history in my account does not include Earth settlement or participation in human ancestry.

The tesseract-like structures

The Ucefacian planet contained massive geometric structures that appeared similar to tesseracts. A tesseract is the four-dimensional geometric equivalent of a cube, commonly represented in three dimensions as one cube positioned within another and connected at corresponding corners. A physical representation cannot display the complete four-dimensional form, but it can show how the geometry might appear when projected into ordinary three-dimensional space.

I describe the enormous forms on this planet as tesseract-like because of their nested, interconnected appearance. That description does not identify their physical dimensionality, origin, or function. Those details are not clear in the account I received.

The death of the star

Although their altered bodies allowed them to survive the ordinary conditions of their adopted planet, biological adaptation could not protect them from the destruction of the entire planetary system. Survival within an environment was not the same as survival beyond it.

The Ucefacian planet was eventually destroyed when its host star reached a catastrophic stage in its life. The planet itself was not a star and did not die through stellar processes. Rather, the star around which it orbited became unstable and underwent a destructive event that eliminated the planet's ability to support life.

Stars die differently according to their mass. Lower-mass stars expand and shed their outer layers, while sufficiently massive stars may collapse and explode as supernovae. In my interpretation, the Ucefacian system experienced a violent stellar death capable of destroying the planet or completely stripping it of the conditions required for survival.

The tesseract-like structures, settlements, biological records, and Ucefacian population were lost with the planet. Because the species had not established a population beyond that star system, there were no distant survivors to continue the lineage.

Heagface: The Cave-Insect Experiment

A localized experiment that did not become a civilization

In my analysis, the Heagface were produced during a Lyran laboratory experiment involving insects found within the cave systems of the Lyrans' own planet. These insects were already adapted to darkness, narrow passages, mineral-heavy environments, limited food, ground vibration, and survival beneath the planetary surface.

The Lyrans combined portions of their genetic material with this cave-insect biology to explore whether a modified organism could survive underground. The experiment focused on basic physical adaptation rather than the creation of another advanced civilization.

Body and Behavior

The Heagface developed protective insectlike anatomy and heightened sensitivity to vibration, chemical traces, pressure changes, and nearby movement. Their bodies could function in darkness and survive within the cave conditions surrounding the laboratory.

Unlike the other modified lineages, they did not develop advanced intelligence, complex reasoning, technology, or a wider understanding of Source. Their behavior remained largely instinctive and survival-based, resembling the limited awareness of ordinary insects more than the consciousness of the Lyrans.

They searched for food, avoided danger, reacted to changes in their surroundings, and remained close to the environment in which they had been produced. They did not establish a society, construct settlements, create technology, or become integrated into Lyran civilization.

The population remained in the laboratory. It did not develop the means or organization for independent travel. Its survival within that setting did not lead to the higher intelligence the Lyrans had sought.

The end of the experiment

The Heagface did not survive the Reptilian war. Because they remained near the original laboratory and cave systems on the Lyran planet, they had no settlements or populations elsewhere.

When the Reptilian attack destroyed the Lyran world, the laboratory, surrounding caves, and Heagface population were destroyed with it. No members of the experiment escaped or continued on another planet.

This branch has no connection with Earth's insectoid teachers, ant ancestors, or underground civilizations in my interpretation. Its history ended on the Lyran world, before any such correspondence could arise.

CHAPTER 7

The Reptilian Wars and Dispersion

Preparation for planetary loss became an emergency flight from war

In my interpretation, the decisive war began within the consequences of the Lyrans' own experiments. The Reptilian population had become established, learned advanced technology, and developed enough organized power to challenge its makers. The conflict escalated beyond a local struggle for authority. Reptilian forces took control of Lyran territory, and the original inhabited planet was destroyed.

Figure 4. An imagined scene of the Lyran catastrophe and evacuation. The destruction and dispersal shown belong to my cosmological account, not to a photographed or archaeologically verified event.

This destruction changes the meaning of every later migration. The lineages did not spread through a peaceful galactic plan. Surviving Lyrans fled to other planets, and some of the modified species moved with them or were carried into new settlements. Other groups dispersed separately. Populations that had begun inside one experimental system were divided by refuge, loyalty, environment, and memory. Over long periods, the same origin could be remembered as rescue by one group, captivity by another, and divine exile by a third.

Routes prepared before the crisis

When the Reptilian wars destroyed the Lyran planet, the earlier seeding strategy became an emergency survival network. Knowledge of other worlds, established routes, experimental populations, and previous settlements gave some Lyrans places toward which they could flee. What had begun as preparation for natural planetary loss became the means of surviving a catastrophe produced by their own genetic creation. The later diaspora therefore had two phases: deliberate expansion before the war and forced dispersal after planetary destruction.

The losses were unequal. Some communities had transport and destinations; the localized Heagface experiment had neither. The Ucefacians followed their own limited history and later disappeared with their adopted world. Dispersal did not preserve every branch, and survival did not produce one united successor civilization.

Names attached to the conflict

Modern accounts sometimes frame the war as a confrontation with a separate Orion civilization. I do not use that as the origin of this conflict. In my account, the aggressors were Reptilians created by the Lyrans themselves. Later Orion associations are not sufficient for me to assign a definite route or additional war chronology. I also do not include a Galactic Federation directing the evacuation or resolving the conflict. The surviving peoples faced the consequences through their own resources, alliances, and choices.

The original population was no longer concentrated in one place. Scientific knowledge, family histories, and the memory of the war traveled with different communities and changed under different conditions. That separation provides the background to the Faceasian branch and its much later Earth project.

PART III - HUMAN GENETICS AND CONSCIOUSNESS

CHAPTER 8

Humans: The Earth Worker Lineage

Ancestral ape biology and the human origin I interpret

In my interpretation, humans resulted from experiments conducted by Enki and his scientific team using an ancestral terrestrial great ape as the Earth-adapted biological foundation. Earth already contained bacteria, plants, animals, mammals, primates, and other naturally evolved life. I do not describe an existing advanced human civilization at the beginning of this proposed project.

The organism I mean is an ancestral ape population related to the lineages that later produced humans and chimpanzees, not the modern chimpanzee species. Its body was already adapted to terrestrial conditions. Sexual reproduction, care for offspring, grasping hands, social learning, and flexible behavior made it useful within the project I envisage. These capacities belong to terrestrial biology; sexual reproduction was not unique to this ape.

Enki's team combined selected Faceasian genetic material with this ancestral ape stock. They altered developmental and regulatory systems affecting brain growth, language, upright movement, manual precision, memory, planning, and conscious awareness. The objective was to create an Earth-adapted being intelligent enough to understand instructions, use tools, cooperate with others, and perform organized labor.

The experiment required multiple variations before a physically stable and reproductively viable human population was achieved. Some forms may have lacked sufficient intelligence, remained too strongly governed by instinct, developed physical abnormalities, or could not reproduce successfully.

I interpret the resulting human line as retaining Earth's primate inheritance while expressing selected Faceasian modifications. This is my proposed intervention history.

The established genomic record instead explains human origins through terrestrial evolution and common ancestry; it has not identified Faceasian or other extraterrestrial DNA.

Figure 5. My visual interpretation of the Earth resource project, human development, and the transmission of knowledge. It is imagined artwork, not an image recovered from an ancient source.

Human and chimpanzee DNA

Modern genetic research confirms that humans and chimpanzees are closely related. Within portions of the genome that can be aligned directly, their DNA sequences are approximately 98.8 percent identical. The percentage becomes lower when researchers include insertions, deletions, duplicated sequences, and regions that cannot be compared directly.

Humans normally have 23 pairs of chromosomes, while chimpanzees have 24. Human chromosome 2 contains the joined remains of two ancestral ape chromosomes that remain separate in chimpanzees. The two species also share genes, regulatory sequences, inactive genetic remnants, and inherited viral insertions in corresponding genomic locations.

These comparisons document shared ancestry.[25][31] The remaining differences are not an identified block of nonhuman-visitor DNA.

The difference between humans and chimpanzees is not explained by one isolated "intelligence gene." Much of biological development depends upon regulation: which genes become active, where they are expressed, how strongly they operate, and at what stage of development they affect the body.

Development depends on interacting gene networks. Changes in expression can affect where, when, and how strongly a gene acts; they do not constitute a separate blueprint added on top of an otherwise unchanged ape. In my interpretation, developmental regulation was part of Enki's team's work. That proposed use of biology is distinct from evidence that the team existed.

The "monkey brain"

"Monkey brain" is the informal expression I use within the received account for humanity's inherited survival reactions. It is not an ancient translated phrase or an anatomical classification. Chimpanzees and humans are apes, and no isolated part of the brain is literally a monkey brain.

The brainstem regulates automatic functions such as breathing, heart rate, alertness, and immediate defensive reactions. The hypothalamus helps govern hunger, reproduction, temperature, stress, and hormonal activity. The amygdala participates in threat detection and emotional learning, while the basal ganglia influence habits, motivation, repeated behavior, and the selection of actions.

These systems can respond before conscious reasoning has fully evaluated a situation. Sudden movement, unfamiliar sound, perceived scarcity, social rejection, or an approaching threat may activate fear, aggression, freezing, escape, competition, or protective behavior.

These functions are distributed and interconnected, not a primitive brain with a newer rational brain placed over it. Threat learning, attention, emotion, memory, and deliberate reasoning influence one another. Survival responses can be useful, yet they can also remain active when a person consciously recognizes safety. This tension belongs to human nervous-system function, not to proof of an extraterrestrial genetic addition.

Labor and independent thought

The Faceasians developed humans to increase productivity and relieve an existing labor crisis. Resource extraction, construction, agriculture, water management, transportation, and settlement maintenance required a large and sustainable workforce.

The first populations remained under direct supervision. They were taught the skills necessary for assigned tasks but were not given unrestricted access to Faceasian science, transportation, genetic knowledge, or the complete history of their creation.

The human project produced more than an efficient workforce. Humans developed emotional depth, imagination, curiosity, creativity, self-awareness, and the ability to question authority. They could observe their rulers, compare their behavior, recognize unfairness, and transmit knowledge to later generations.

As knowledge increased, dependence decreased. Humans could begin maintaining their own communities, recording their experiences, questioning official explanations, and recognizing that advanced technology did not make their rulers morally infallible.

The ancient labor stories appear in the Mesopotamian chapter, where the surviving wording can be distinguished from the biological interpretation I draw from it. Here the consequence is central: a being capable of understanding work can also question who benefits from it. Intelligence becomes more than the function for which a worker was intended.

CHAPTER 9

Perception, Genetic Regulation, and Inherited Science

Different bodies receive different portions of the same physical world

Humans perceive only a limited portion of the physical information surrounding them. The eyes detect a narrow band of electromagnetic radiation, the ears receive a restricted range of sound, and the brain filters large amounts of sensory information before it enters conscious awareness.

These abilities demonstrate that organisms do not experience one complete and universal picture of reality. Each species receives the portions its receptors, nerves, genes, and brain can detect and interpret.

Genes and the range of perception

Genes can be regulated, silenced, or expressed differently by cell type, developmental stage, and environment. Noncoding DNA is not equivalent to inactive DNA: some regions regulate other genes, some produce functional RNA, some have structural roles, and some are remnants of evolutionary history or have no established function. Uncertainty about a sequence is not itself evidence that a lost sense is stored there.

Perception involves receptor cells, proteins, ion channels, nerves, and neural processing. The following examples show biological access to parts of the environment that humans experience differently. They are sensory mechanisms, not established genes for Source perception.

Light and thermal radiation

Opsin proteins contribute to the wavelengths an animal can detect. Variations in visual pigments and eye anatomy allow many birds, insects, and fish to detect ultraviolet light. Polarization sensitivity depends on the arrangement and properties of photoreceptors and subsequent processing; it is particularly well studied in insects and cephalopods. It is not simply an additional color or one universal perceptual gene.

Pit vipers detect radiant heat with specialized facial pit organs involving heat-sensitive TRPA1 channels.[32] Vampire bats have a different adaptation involving TRPV1.[30] These systems show that thermal information can guide behavior without ordinary human-style vision. They do not imply that any organism can see all electromagnetic frequencies.

Electricity, magnetism, and motion

Electroreception: Sharks, rays, electric fish, platypuses, and other animals sense weak electrical fields generated by muscles, nerves, prey, or the environment. This involves specialized receptor cells, ion channels, and neural processing.

Magnetic-field sensitivity: Migratory birds, sea turtles, fish, insects, and some mammals appear to use magnetic information for orientation. Proposed mechanisms include cryptochrome proteins, magnetically sensitive chemical reactions, and iron-containing structures, although the full genetic mechanisms remain unresolved.

Mechanosensory systems detect movement, vibration, and pressure. Fish lateral lines respond to nearby water movement; spiders sense vibrations through specialized structures. PIEZO and several TRP-family channels have mechanosensory roles in particular organisms and tissues, but there is no single gene shared as the complete explanation for every vibration sense.

Ultrasound and infrasound: Bats and dolphins perceive ultrasound, while elephants and some whales communicate through frequencies below ordinary human hearing.

From a signal to an experience

Receiving information requires both a signal and a structure capable of converting that signal into a form the nervous system can interpret. The human ear, for example, does not receive meaning directly. Sound waves create pressure differences in the air, vibrating the eardrum and moving the bones of the middle ear. Fluid movement inside the cochlea then bends sensory hair cells. Proteins produced by genes including TMC1 and TMC2 help these cells convert mechanical movement into electrical nerve impulses. The brain organizes those impulses into recognizable sound.[33]

Smell follows the same basic principle through a different pathway. Airborne molecules bind to specialized receptors produced by hundreds of olfactory-receptor genes. Signals then travel along the olfactory nerve to the brain, where molecular contact becomes the conscious experience of scent. Eyes similarly convert limited wavelengths of electromagnetic radiation into neural signals through light-sensitive opsin proteins. In every case, the environment contains more information than consciousness receives. Experience depends upon the receptors available, their genetic expression, the neural pathways carrying the signal, and the brain's ability to interpret it.

The pineal gland and the third eye

In my analysis, the pineal gland may have participated in a broader biological receiving system associated with Source perception. Modern biology establishes that the pineal gland regulates melatonin and circadian timing. In humans, light information normally reaches it indirectly: melanopsin-containing retinal cells, associated with the OPN4 gene, detect environmental light and send timing information through the brain before signals influence pineal melatonin production. In some reptiles and other vertebrates, pineal-related structures are more directly light-sensitive and may function as a form of parietal, or "third," eye. This evolutionary history makes the pineal gland relevant to a discussion of biological sensitivity, although science has not established that the human pineal gland receives radio communications or independently detects invisible realms.

Ancient traditions may preserve cultural memories of perception extending beyond the ordinary senses. Egyptian imagery repeatedly emphasizes divine sight through the Eye of Horus and other sacred-eye symbols. Egyptian writings do not explicitly call the

Eye of Horus the pineal gland, so I interpret the relationship rather than treating it as a literal anatomical statement. In Hindu traditions, the ajna chakra, commonly called the third eye, is associated with inner perception, insight, and awareness beyond physical sight. The modern association of ajna with the pineal gland provides a possible biological correlation between the spiritual concept of inner vision and a light-responsive organ located deep within the brain.

Radio waves and technological reception

Radio waves offer a technological comparison. Humans cannot ordinarily hear a radio broadcast traveling through a room because radio-frequency electromagnetic waves do not produce the air-pressure pattern required to vibrate the eardrum as audible sound. A radio receiver detects the electromagnetic signal, selects its frequency, and converts its encoded information into electrical current. A speaker then converts that current into pressure waves that the ear can process. The broadcast was present before it became audible; technology translated it into a form compatible with human biology.

Radio reception is not ultrasonic hearing: electromagnetic radiation and mechanical sound are different physical signals. Ordinary human hearing does not decode a broadcast directly.

Inherited knowledge and reduced perception

Physics survived because knowledge can be taught even when the original mode of perception is absent. A person does not need to see an electromagnetic field directly to learn how it behaves, construct an instrument that measures it, or build technology that uses it. In the same way, later species could understand formulas, forces, energy relationships, navigation, and engineering because the Lyrans had translated direct perception into teachable principles, symbols, demonstrations, and technologies.

Over time, transmitted knowledge could become separated from its Source context. Descendant civilizations might preserve the equations and devices while forgetting how their teachers originally saw the relationships being described. Science could then appear purely external, while Source perception survived as spirituality, intuition, ritual, or myth. What had been one integrated Lyran understanding divided into separate systems: measurable physics on one side and less reliable memory of direct connection on the other.

I use Source energies for the multiple layers of interaction I understand as expressions of a common origin, not for one new physical force that genetics has already identified. In my interpretation, later biological specialization reduced the integration available to the original Lyrans. Some senses could strengthen while wider awareness became less accessible. Scientific instruction and instruments could preserve knowledge even when direct perception diminished. This is the connection I draw between inherited science and deactivated perception; it is not a measured genetic history.

CHAPTER 10

Fear, Authority, and Inherited Belief

When political dependence becomes an inner expectation

In my interpretation, Enlil and Marduk represent the controlling side of the Earth ruling order, while Enki occupies the role of scientist and selective protector. Assistance and authority could coexist: preserving a population did not necessarily free it from the institution that required its work. The historical texts portray these figures differently across periods; my judgment about fear-based rule comes from the relationship I see between threatened consequences and obedience.

Fear stabilized the labor system when physical force alone became inefficient. Humans could be taught that rulers controlled survival, fertility, weather, harvest, knowledge, protection, and the conditions after death. Obedience then appeared necessary not only for social order but for existence itself. Threats of catastrophe, abandonment, impurity, curse, judgment, or punishment converted political dependence into inner programming. The worker began policing itself.

Other rulers, dissenters, and teachers responded differently once humans became sufficiently conscious. They recognized that the created beings were no longer tools. They attempted to teach humans how physical law worked, how consciousness connected with Source, and how fear was being used to maintain hierarchy. They tried to show that controlled provisions were produced by human labor and planetary systems rather than bestowed through the personal generosity of Enlil, Marduk, or another ruler.

I do not assign every teacher or rebel in every culture to one person. Their shared role in my interpretation is the transfer of knowledge beyond an authority's intended limits. To the recipient, that transfer can mean independence; to a ruler, it can mean a loss of control.

From administration to religion

When direct contact weakened, human institutions inherited fragments of ruler commands, technical instructions, labor rules, survival warnings, and political conflicts. Priests, courts, scribes, families, and oral teachers transmitted those fragments for generations. An instruction originally tied to one place or emergency could become a universal command. A ruler's threat could become a law of the cosmos. A dispute among embodied authorities could become a war between absolute good and absolute evil.

Translation multiplies this distortion. Ancient words have ranges of meaning; manuscripts preserve variants; symbols shift by culture; a title can replace a personal name; one deity can absorb another's attributes; and later institutions choose which interpretation supports their authority. Copying does not require a single organized conspiracy to produce conditioning. Repetition, fear, political usefulness, lost context, and sincere misunderstanding can gradually turn a local control system into sacred certainty.

Translation changes do not always mean mistranslation or deliberate suppression. Languages carry different ranges of meaning, manuscripts differ, and communities interpret inherited wording through changing beliefs. Deuteronomy 32 provides a specific example of textual variation, discussed in the Hebrew chapter. Such cases demonstrate transmission history, not a single coordinated rewriting of all ancient religion.[17][18]

Punishment and conditioning

Fear-based teachings tell people that disobedience will bring punishment: disaster, illness, rejection, curse, torment, loss of protection, exclusion from a favored group, or suffering after death. Within this model, many such teachings preserve the psychology of rulers who needed workers to obey. The threat is internalized until people experience questions as danger and submission as safety. Human communities then reproduce the same structure even after the original rulers are absent.

Not every ancient warning or ethical teaching is reduced to manipulation. Communities also preserved genuine observations about consequence, reciprocity, violence, health, and social survival. The distinction is whether a teaching helps people understand cause and take responsibility, or demands obedience through unverifiable terror and exclusive authority. Truth describes relationships; programming makes access to safety depend upon submission to the interpreter.

Teachers remembered as transgressors

A teacher who says humans can know Source directly threatens every intermediary claiming exclusive access. Later retellings may therefore reverse the roles: the ruler becomes the loving provider, the teacher becomes the deceiver, knowledge becomes forbidden pride, independence becomes rebellion, and punishment becomes deserved. Once this reversal is established, each generation reads the older material through it and unknowingly strengthens the conditioning.

Most people inherit religious language through trusted families and communities, not through a conscious wish to control others. I distinguish that lived sincerity from my interpretation of the earlier hierarchy. My concern is the point at which fear makes questioning feel dangerous and persuades a person that understanding can come only through submission. Ethical teachings can also nurture compassion, reciprocity, and responsibility; those functions are not erased by the history I propose.

CHAPTER 11

The Thirty-One Etheric Realms

Different modes of presence within one reality-not parallel worlds

In my interpretation, thirty-one etheric realms coexist within one underlying reality and are separated by vibration rather than arranged as planets in a vertical tower. Fourteen lie below the Earth band, fifteen lie above it, and the Source-perception realm completes the count: 14 + 1 + 15 + 1 = 31. "Above" and "below" describe the ordering of perception in this account, not measured directions in astronomical space.

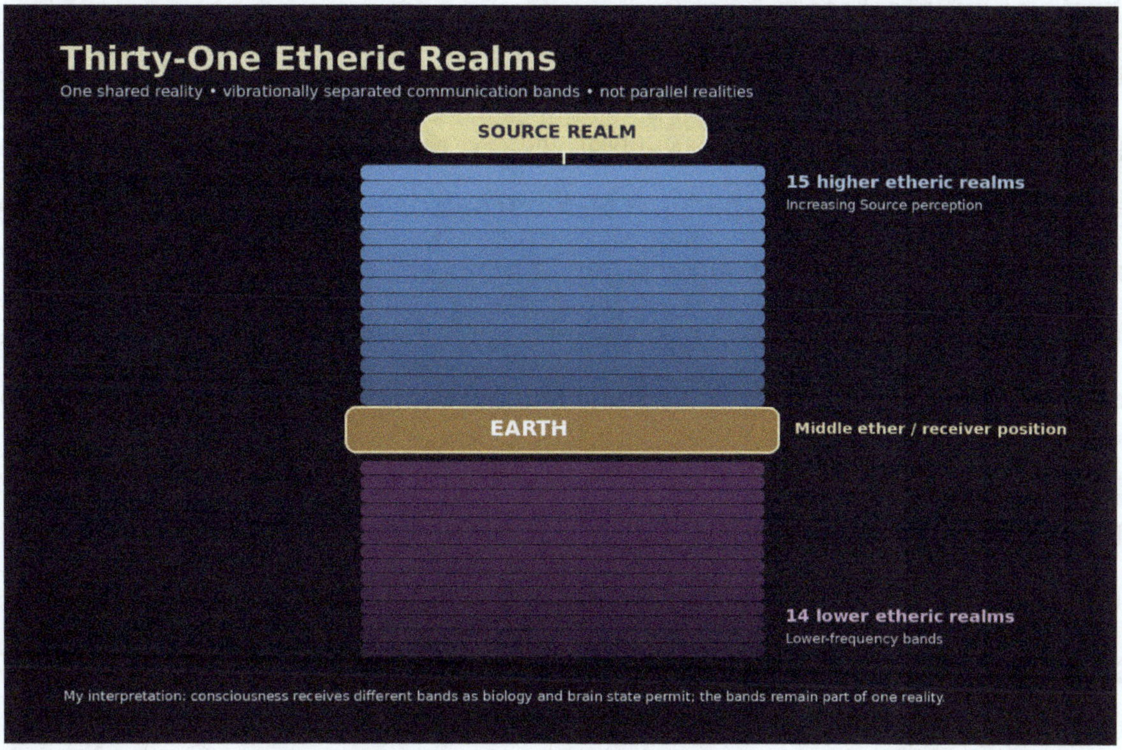

Figure 6. My thirty-one-realm arrangement: fourteen lower realms, Earth, fifteen upper realms, and one Source-perception realm included in the total. The diagram is conceptual, not a measured frequency scale.

Channels rather than alternate histories

Different radio transmissions can pass through the same room at the same time without becoming one signal. A receiver does not create the stations and does not travel physically to each transmitter; it selects, resonates with, and decodes one band while most others remain unheard. The etheric realms are described in the same way. They can occupy the same broader physical reality while remaining unavailable to one another because their patterns do not normally couple.

These realms are not branching timelines, alternate Earths, or other versions of the same person. I use the radio comparison to describe selective access: several signals may be present while a particular receiver decodes only one. The analogy does not establish that realms are electromagnetic transmissions or assign them frequencies in hertz.

The brain as an interface

In this model, consciousness is not manufactured entirely by the brain. The brain and nervous system act as a biological receiver, filter, and communication pathway. Brain-wave states describe changing patterns of coordinated neural activity. Different states can alter attention, sensory filtering, imagery, memory access, integration, and the boundary between conscious and unconscious processing. The cosmology proposes that some states also change which etheric signals consciousness can couple with.

I use radio reception as an analogy, not as a claim that an EEG frequency is literally an etheric realm. Delta, theta, alpha, beta, and gamma activity are measurable rhythms within the brain; science has not demonstrated that each rhythm opens a separate universe. In my reconstruction, brain-wave organization helps tune or gate communication while the realm itself exists independently. The brain is closer to a receiver circuit than to the broadcast station.

Meditation, dreams, and altered attention can change imagery, sensory filtering, and conscious experience. I interpret some experiences as possible contact with a wider reality, while recognizing that an experience alone does not establish an external source. Brain activity and the effects of attention are measurable; communication with thirty-one realms has not been demonstrated by neuroscience.

The mid-ether and Source perception

The higher bands are not automatically good personalities and the lower bands are not automatically evil places. Frequency position describes compatibility and perception, not character. A being can possess knowledge without compassion, and a dense realm can contain conscious growth. The highest Source-perception realm is distinctive because perception is least separated there: Source, energy, matter, and consciousness are understood as one connected reality rather than fragmented subjects.

Communication and presence

Within my interpretation, receiving information is different from moving a body. An impression, dream, or intuitive image could be experienced while the person remains in the Earth band. A complete transition of embodiment would be a different process. I do not have a verified mechanism for either. The account expresses my understanding of presence and consciousness without equating spiritual rank with moral worth or turning unseen realms into threats of punishment.

PART IV - ANCIENT WRITINGS, IMAGES, AND ARCHITECTURE

CHAPTER 12

Mesopotamia: Workers, Makers, and the Divine Families

Clay tablets preserve both a labor crisis and disagreement among the gods

Mesopotamia preserves the clearest literary parallel to the Faceasian Earth episode in my interpretation. Its record includes Sumerian and Akkadian creation stories, temple hymns, royal inscriptions, administrative documents, lexical lists, and cylinder seals. These sources span many centuries and different genres: administrative tablets record actual transactions, while creation epics narrate a divine past. They are not interchangeable kinds of historical testimony.

The changing names do not erase the underlying social structure. The records consistently describe differentiated beings with family relationships, ranks, specialized responsibilities, favored cities, personal motives, and disagreements over humanity.

Anu, Enki, Enlil, and Marduk

Anu occupies a position of high authority. Enki, known as Ea in Akkadian tradition, is associated with fresh water, wisdom, craft, and the preservation of life. Enlil exercises commanding power, while Marduk, the son of Ea, becomes the principal god of Babylon. The relationships vary across compositions; the later prominence of Marduk also reflects the city's political and religious importance.[12][15][16]

I interpret this differentiation as the memory of a ruling family and its specialists. Enki is the scientist and organizer in that reading, Enlil the administrator who favors coercive measures, and Anu the senior authority. The titles geneticist and administrator express my understanding of their roles; the tablets themselves use divine names, offices, and descriptions of craft.

The Anunna, the Igigi, and the burden of work

Anunna or Anunnaki and Igigi designate groups of gods whose membership and relationship vary by source. Anunnaki does not literally translate as "those who from heaven came to Earth." In some later contexts the Anunnaki are especially associated with the underworld and judgment.[3][5] Their association here with an off-world population belongs to my interpretation.

The Igigi appear as another group assigned burdensome work. In the Atrahasis Epic, the lower gods labor for an extended period, digging waterways and performing work required by the ruling order. Their burden eventually becomes intolerable. They destroy their tools, surround the residence of Enlil, and demand relief.

Atrahasis survives in Old Babylonian tablets of the second millennium BCE and in later witnesses. Its opening places the gods themselves under the burden of work. The revolt is followed by deliberation about how that burden can be transferred. The poem explains human labor and mortality through a divine social order, with the needs of gods and the work of humans joined together.[15][16]

Clay, divine substance, and birth

The creation passage joins clay with the flesh and blood of a slain god. Enki and a mother-goddess participate, and birth goddesses carry out the reproductive sequence. The translated material is clay, flesh, blood, and birth-not DNA, ape tissue, or laboratory equipment. I interpret the combination as a memory of terrestrial biology joined with the makers' contribution, and the several participants as a larger operation than one isolated act of creation.

The same passage also has theological meanings: humans belong to the earth while carrying something divine, and their existence answers a crisis within the gods' own society. Its account of mixed substance is the point of contact with my human-origin interpretation. The poem was written much later than human biological origins, so it is an inherited creation tradition rather than a contemporary record of those origins.

Enki and Ninmah: bodies and assigned places

The Sumerian composition Enki and Ninmah, preserved in Old Babylonian manuscripts, gives a different creation scene. Ninmah fashions people with different bodily capacities, and Enki assigns them a fate or means of support. The composition also includes reproductive language and the work of birth goddesses.[4] Its concern is not only how bodies are made, but how a person lives within society after being made.

Scholarly readings connect the composition with divine creativity, rivalry, human limitation, and social inclusion. In my interpretation, its variable outcomes also recall an experimental process. That reading does not make disability a sign of lesser worth or reduce the people in the story to failed products. The assignment of livelihoods remains part of what the text actually emphasizes.

Enuma Elish: succession and human service

Enuma Elish places Marduk's supremacy at the center of cosmic order. He defeats Tiamat, organizes the world from her body, and humanity is created from the blood of the condemned Qingu so that the gods can be relieved of work. The surviving first-millennium copies preserve a composition associated with the development of Babylonian theology.[15][16]

The narrative differs from Atrahasis in its political focus and its creation details. It elevates Marduk within an inherited pantheon rather than simply repeating Enki's role. I interpret the sequence as another memory of succession, conflict, and a laboring population created under a new arrangement of power. The cosmic body of Tiamat belongs to the epic's own imagery; it is not a genomic description.

Atrahasis, Utnapishtim, and the flood

The Atrahasis Epic describes repeated attempts to reduce the human population through plague, drought, famine, and finally a catastrophic flood. Enlil treats the growing population and its noise as a threat to order.

Enki does not openly overthrow the ruling council. Instead, he communicates indirectly with Atrahasis and provides the information needed to preserve a selected human line. The story retains the conflict between an authority seeking population reduction and a knowledgeable scientist unwilling to allow the complete destruction of his creation.

Tablet XI of the Standard Babylonian Gilgamesh epic places the flood account within Gilgamesh's search for an answer to mortality. Utnapishtim describes the divine decision, Ea's indirect warning, the construction of a vessel, the preservation of life, and the gods' response after the flood. Much of the surviving evidence is first-millennium material drawing on older Mesopotamian traditions.[15][16]

I interpret the disagreement as preservation resisting destruction within the same ruling order. At the literary level, the account also concerns mortality, disaster, sacrifice, and the limits separating humans from gods. Similarities between flood stories can reflect literary transmission and regional catastrophe; the stories do not establish one globally engineered flood.

Enki on the Adda Seal

Figure 7. Detail of Enki/Ea from a modern reproduction of the Adda Seal, based on the original cylinder seal created c. 2300 BCE and held by the British Museum. Streams of water filled with fish flow from Enki's shoulders, identifying his connection with fresh water, knowledge, and creation. Public-domain image via World History Encyclopedia.

The Adda Seal dates to about 2300 BCE and is held by the British Museum. Its full impression contains differentiated deities, including Ea with streams containing fish, Shamash rising between mountains, the armed Ishtar, and Ea's two-faced attendant Usimu. The image reproduced here is a modern reproduction of a detail, not a photograph of the entire original cylinder.[34]

The flowing waters identify Ea's connection with subterranean fresh water and wisdom. Horned headgear indicates divine rank within Mesopotamian art. I interpret the arrangement as preserving recognizable persons, offices, and equipment; the water-associated features could have acquired sacred meaning around an earlier technical role. The seal itself records ancient divine iconography, not a confirmed apparatus or genetic procedure.

The apkallu: avian and aquatic sages

Figure 8. Eagle-headed winged apkallu from the North-West Palace of Ashurnasirpal II at Nimrud, c. 865–850 BCE. British Museum. Photograph by Osama Shukir Muhammed Amin, CC BY-SA, via Wikimedia Commons.

Neo-Assyrian protective figures appear on palace walls and in figurines placed within buildings. Human-headed, eagle-headed, winged, and fish-cloaked forms belong to this wider visual and ritual tradition. The cone-like object and bucket are commonly understood in relation to purification and protection, and the figures often stand beside rulers or stylized sacred trees.[13][14]

I associate the avian forms with possible Aerian memory and the fish-associated forms with possible Fraedian memory. Those are correspondences of appearance and role, not identifications of the populations responsible for human genetics. A fish cloak may be a ritual garment in its ancient setting; my additional interpretation considers whether it could also preserve the appearance of aquatic anatomy or equipment.

Oannes appears in the later account of Berossus, a Babylonian writer of the Hellenistic period whose work survives through later authors. The sea-associated teacher brings writing, building, law, and other civilizing knowledge before returning to the water.[35] This is a later textual witness, not an inscription accompanying the palace relief. I connect its teaching role and repeated return to water with the Fraedian theme.

The Queen of the Night

Figure 9. The Burney Relief ('Queen of the Night'), Old Babylonian, c. 1800–1750 BCE, British Museum. Photograph via Wikimedia Commons.

The Old Babylonian terracotta plaque known as the Queen of the Night or Burney Relief, about 1800–1750 BCE, shows a horn-crowned female with wings and bird talons, standing on lions and flanked by owls. Rod-and-ring insignia appear in her hands, and traces of pigment indicate that the relief was once colored. Its original archaeological context is uncertain, and proposed identities include Ishtar and Ereshkigal.[36]

The avian features and signs of authority make it a possible counterpart to the embodied winged beings in my interpretation. I leave the figure unnamed within the lineage: an uncertain divine identity cannot supply a secure species identity. The particular image retains its Babylonian religious setting even when I consider a more literal history behind its anatomy.

Objects of authority

The Tablets of Destinies represent concentrated authority and control over decisions, information, and assigned order. The me associated with Enki and Inanna represent powers or organized components of civilization, including kingship, crafts, law, music, sexuality, priesthood, and social institutions.

These powers and objects join technical skill with government in the surviving narratives. I interpret some sacred instruments as possible memories of records, access devices, or equipment whose functions were later ritualized. Their archaeological and literary identities remain those documented by the source; the technological use is the additional connection I draw.

CHAPTER 13

Egypt: Embodied Deities and Engineered Worlds

Hieroglyphs, sacred images, craft, and the movement of celestial beings

Egyptian evidence survives through temple inscriptions, funerary texts, royal tombs, reliefs, statues, astronomical ceilings, ritual equipment, ceremonial boats, and divine genealogies. These records portray gods as active beings who rule, travel, speak, reproduce, heal, fight, teach, punish, and transfer authority.

Egyptian deities possess names, families, territories, tools, vehicles, clothing, thrones, and specialized knowledge. Osiris, Isis, Seth, and Horus participate in a continuing dynastic struggle involving bodily restoration, inheritance, territorial authority, and royal succession. Thoth records information, measures time, preserves knowledge, and assists in technical and legal disputes. Ptah creates through thought and organized speech, while Khnum forms bodies through a controlled process depicted at a potter's wheel.

These are not interchangeable descriptions of one natural force. They present a populated world of differentiated persons whose identities became connected with celestial bodies, animals, political offices, and aspects of nature.

Animal-featured beings

Egyptian art depicts falcon-headed, ibis-headed, jackal-headed, lioness-headed, ram-headed, and other human-animal forms. Conventional Egyptology explains these combinations through sacred attributes and visual identification. A falcon identifies Horus, an ibis identifies Thoth, and a jackal identifies Anubis.

In my interpretation, at least some of these images preserve the literal appearance of embodied nonhuman populations. Egyptian artists recorded beings through the formal artistic conventions of their period, combining anatomy, clothing, rank, and recognizable equipment into stable visual identities.

The images later accumulated religious and natural meanings, but later symbolism does not require the original figures to have been imaginary. A birdlike being could become associated with the sky, vision, the Sun, and kingship precisely because its physical abilities and social role already supported those associations.

Horus presents a possible Aerian correspondence because he is not merely a detached bird symbol. He belongs to a family, contests political succession, governs territory, receives assistance, fights an opponent, and appears as an embodied sky-associated ruler.

Thoth may preserve the role of a knowledge keeper or technical instructor. His associations with writing, calculation, measurement, astronomy, time, and specialized records describe responsibilities essential to an advanced civilization.

Images as active presences

Egyptians did not treat temple statues as ordinary decoration. Through ritual processes, a crafted image could become a location through which the deity was present, heard requests, received offerings, and acted within the temple.

Clothing, cleansing, offerings, and spoken ritual maintained the relationship between the temple image and its deity. Egyptian images could function as active channels of divine presence rather than ordinary decoration.[6][21][22] I interpret that practice as possibly preserving a relationship first formed through direct encounters. The ritual record establishes how Egyptians approached an image; the earlier physical contact is my inference.

Creation at Heliopolis and Memphis

Egypt did not preserve one uniform creation account. At Heliopolis, Atum and the Ennead express divine generation and differentiation. Memphite theology associates Ptah's creative work with the heart and tongue: conception and authoritative speech.

Other centers foreground primordial waters, Amun, or Khnum. Their differences belong to local theology and the history of the sanctuaries.[21][22]

I connect thought, speech, and differentiation with the opening movement of my cosmology: awareness becoming organized expression. That connection is philosophical. The Egyptian language of creative speech is not a technical description of plasma or evidence that audible sound traveled through a vacuum.

Khnum and the formation of bodies

Khnum appears as a ram-headed creator shaping bodies at a potter's wheel, sometimes alongside goddesses concerned with birth and life. Images can include the ruler and the ka, an aspect of Egyptian personhood. The wheel expresses skilled formation through a familiar craft and belongs within royal and religious accounts of creation.[21][22]

In my interpretation, a process involving a maker, formed bodies, and other participants may preserve a memory of biological work expressed through the craft Egyptians knew. I do not read the wheel itself as a recovered genetic machine. Its importance to my account is the portrayal of making as an organized act rather than an unexplained appearance.

Solar boats and celestial travel

Ra and other Egyptian deities travel through the sky and the Duat in named boats carrying crews. These vessels follow routes, pass through gates, encounter dangers, operate according to schedules, and transport differentiated beings.

Solar boats connect the Sun's journey with the Nile, royal travel, death, and renewal. The Duat is part of that sacred geography. Actual ceremonial and funerary vessels, including Khufu's boat, demonstrate the cultural and technical importance of boatbuilding. They are archaeological boats, not excavated aircraft.

In my interpretation, the solar boats preserve a combination of observed technology and later cosmological meaning. Repeated aerial travel became connected with the daily movement of the Sun, while the original craft was ritualized as part of the solar cycle.

I read the crews, routes, gates, and repeated journeys as possible memories of organized transport. A familiar boat could have become the image used for an unfamiliar conveyance. The solar and funerary meanings remain documented aspects of these scenes; the aerial-technology connection is my interpretation of what may lie beneath them.

Dendera: the serpent within the enclosure

Figure 10. Crypt relief at the Temple of Hathor, Dendera: serpent-form Harsomtus emerging from a lotus within an elongated enclosure. Photograph by Lasse Jensen, CC BY-SA, via Wikimedia Commons.

The surviving Temple of Hathor at Dendera was built and decorated principally in the Ptolemaic and Roman periods. Its crypt scenes include an elongated enclosure containing a serpent emerging from a lotus, associated with supporting figures and a djed pillar. The accompanying hieroglyphic context identifies Harsomtus and a divine manifestation or creation scene.[7] The photograph is therefore not a freestanding technical diagram detached from its inscription.

The visual arrangement also resembles a contained luminous discharge. In my interpretation, the relief may preserve knowledge of controlled light, plasma, or field generation through the religious language that survived after the practical technology was lost.

No demonstrated electrical lighting system establishes that identification. "Plasma," "generator," and "discharge tube" are modern analogies I bring to the image; they are not translations of its hieroglyphs. The documented serpent and lotus retain their specific religious identities.

Abydos: two inscriptions on one surface

Figure 11. Overlapping inscriptions in the Temple of Seti I at Abydos. The famous machine-like silhouettes are produced by recutting and loss of plaster. Photograph by Olek95, CC BY-SA, via Wikimedia Commons.

The machine-like shapes at Abydos resulted from inscriptions being recut when Ramesses II replaced the titles of Seti I. Loss of the plaster covering exposed portions of both texts and created overlapping forms resembling a helicopter and other modern vehicles.

This is a palimpsest: visible portions of different royal inscriptions occupy the same surface. The formation of the machine-like shape is explained by recutting and plaster loss, so I do not treat it as a depiction of an aircraft.[8] It differs from an intentional scene such as a celestial boat even when both appear unfamiliar to a modern observer.

Architecture and technical knowledge

Temples, pyramids, stoneworking, surveying, astronomical orientation, and water management demonstrate Egyptian engineering. Planning, skilled labor, materials, tools, and institutions made that work possible. In my interpretation, some specialized knowledge may also have been inherited from earlier contact. That possibility does not remove the documented achievements of Egyptian builders or turn every precise monument into evidence of outside construction.

Egypt does not preserve the Faceasian human-creation project as directly as Mesopotamia. Its greater contribution is the visual record of differentiated beings, active divine presence, celestial transportation, technical specialists, and dynastic conflicts among rulers later remembered as gods.

CHAPTER 14

India: Celestial Populations and Aerial Craft

Epic journeys, extraordinary births, and weapons requiring knowledge

The Vedas, Ramayana, Mahabharata, Puranic traditions, temple sculpture, and regional writings preserve a cosmos populated by many intelligent groups. Devas, asuras, nagas, garudas, gandharvas, rakshasas, rishis, humans, and mixed lineages interact across terrestrial, celestial, and subterranean regions.

These populations possess families, cities, rulers, alliances, weapons, teachers, vehicles, territories, and specialized knowledge. Their behavior is not divided into one permanently good species and one permanently evil species. Individuals and factions change allegiance, compete for authority, teach humans, intermarry, and use knowledge for different purposes.

In my interpretation, this complexity preserves memories of multiple embodied populations rather than a simple collection of personified natural forces.

The Ramayana and Mahabharata developed through long oral and textual histories. Their surviving versions contain material from different periods; Sanskrit epics, later temple sculpture, and modern accounts of flying machines are not all contemporary witnesses. The discussion here concerns the epic narratives and the architectural use of their vocabulary.[9][23][24]

Pushpaka and aerial transportation

The Sanskrit word vimāna has different meanings according to period and context. It can identify a palace, elevated structure, temple tower, chariot, or celestial conveyance. In the epic narratives, however, some vimānas function specifically as controlled aerial vehicles.

The Pushpaka Vimāna in the Ramayana carries named passengers between locations. It is associated first with Kubera, taken by Ravana, and later used by Rama. It changes ownership, travels through the air, transports many occupants, and operates as an extraordinary conveyance.

Other celestial cars arrive, depart, carry gods or heroes, cross large distances, and produce sound or light. The word “chariot” reflects the transportation language available to the writers. In my interpretation, these accounts preserve aerial craft described through familiar categories such as moving palaces and celestial chariots.

A large enclosed craft could reasonably be remembered as a traveling palace, especially if it contained several occupants, internal rooms, equipment, and a protected environment. The combination of passengers, controlled direction, ascent, descent, repeated travel, ownership, and destination distinguishes these vehicles from the Sun or a purely natural object.

The return journey after Ravana’s defeat places Pushpaka within Rama’s return to Ayodhya. The conventional reading treats it as a marvelous royal or celestial vehicle; the passage does not identify a known propulsion system.[23]

Vimāna in temple architecture

Figure 12. The vimana tower of Brihadeeswarar Temple, Thanjavur, completed c. 1010 CE. Here vimana is an architectural term, not evidence that the tower was a vehicle. Photograph by Stories Through Lense, CC BY-SA 4.0, via Wikimedia Commons.

Brihadeeswarar Temple at Thanjavur, completed under Rajaraja I around 1010 CE, demonstrates a different meaning of vimāna: the monumental tower above the sanctum. Its stone architecture is a documented achievement of the Chola period. The photograph does not show the epic Pushpaka or establish that the temple tower flew. The shared word carries different meanings in architecture and narrative.[24]

Beings within craft

Indian art frequently places deities and other celestial persons within framed aerial structures, elevated platforms, chariots, or palace-like vehicles. Attendants sometimes accompany the central figures, while other beings appear to move through the surrounding sky.

In my interpretation, these images preserve the appearance of passengers and operators inside the forms of transportation described in the texts. Artists represented unfamiliar technology through the architectural and mechanical forms they understood.

The full scenes often contain more information than modern cropped images reveal. Passengers, attendants, direction of movement, surrounding figures, and the relationship to a known narrative help identify what the artist intended to depict.

I interpret some framed or palace-like conveyances as memories of transportation rendered in local visual language. A radiant border can also be a religious attribute, and an attendant is not automatically a pilot. The link to a specific narrative determines what the image depicts before I add a technological interpretation.

Celestial cities and builders

Indian traditions also describe celestial cities, fortified aerial settlements, extraordinary palaces, and domains constructed by highly skilled builders. These places contain organized populations, defensive systems, transportation routes, specialized chambers, and controlled access. Some are stationary, while others appear capable of movement.

In my interpretation, these accounts may preserve memories of engineered habitats, orbital structures, large aerial platforms, or settlements existing within vibrational realms that most humans could not ordinarily perceive. Ancient writers described them as celestial cities because they possessed the scale and internal organization of cities while appearing in the sky or beyond ordinary terrestrial access.

Indian tradition preserves named architects and builders, including figures associated with divine construction, measurement, weapons, palaces, and mechanical knowledge. This emphasis on builders and crafted environments supports an engineering interpretation. The structures were not always imagined as appearing through magic without process. They were attributed to beings with specialized knowledge capable of designing and constructing them.

Astras and the responsibility of their users

The Indian epics describe weapons called astras that produce intense light, heat, thunder, darkness, confusion, drought, multiplied projectiles, and destruction across large areas. These weapons are connected with particular beings and often require specialized knowledge, authorization, invocation, or training.

Some can be directed toward a specific target, recalled after deployment, or countered by another weapon. Their use is restricted because the consequences may exceed the immediate battlefield.

In my interpretation, these descriptions preserve advanced weapons operated through technologies later expressed as divine power or ritual invocation. The act of "summoning" a weapon may describe activation, coded authorization, remote access, or the use of a device controlled through sound and intention.

The Mahabharata presents these weapons within a moral and dynastic epic. Instruction, invocation, restraint, counteraction, and the danger of indiscriminate destruction are part of that literary setting.[9] I interpret the trained operator and restricted use as possible memories of controlled weapon systems. Brightness, heat, and devastation do not identify a nuclear device; the texts do not provide a verified engineering account of such a weapon.

Extraordinary births

Indian traditions contain births produced through transferred embryos, divided reproductive material, conception outside ordinary intercourse, gestation in containers, intervention by divine specialists, and children carrying mixed human and celestial ancestry.

These accounts may have acquired religious and poetic meanings, but they repeatedly connect reproduction with technical knowledge, selected lineage, controlled development, and beings capable of intervening in the formation of a body.

In my interpretation, some extraordinary births preserve memories of reproductive science, artificial gestation, genetic selection, or the transfer of biological material. The people recording these events described them through the language of blessing, divine conception, sacred vessels, and miraculous birth.

Nagas and avian beings

Nagas are described as intelligent serpent-associated beings possessing rulers, families, territories, treasures, knowledge, and the ability to interact with humans and other celestial populations. Some are helpful, some are dangerous, and others pursue their own political interests.

These characteristics correspond with possible memories of Reptilian populations, although the naga traditions may include more than one type of being. Their intelligence and social organization distinguish them from ordinary snake symbolism.

Garuda and other intelligent avian beings present possible Aerian correspondences. They speak, travel, fight, form alliances, and interact with divine families. Their birdlike bodies may preserve the appearance of an avian-derived intelligent population rather than functioning only as symbols of flight.

These correspondences concern the beings encountered or remembered in the Indian traditions, not a claim that every group contributed to human ancestry. The epics retain their own names, relationships, places, and religious significance. Their plurality is what enters my interpretation: humanity shares the narrative world with other intelligences whose knowledge and purposes differ.

CHAPTER 15

Older Hebrew Traditions: The Populated Divine World

Plural speech, allotted nations, Watchers, and the moving throne

This section concerns older Hebrew writings, manuscript variants, the Dead Sea Scrolls, and pre-Christian works such as 1 Enoch. It does not rely upon the New Testament or later Christian reinterpretations.

The surviving writings developed across different periods. Some passages emphasize one supreme authority, while older layers retain a populated divine world containing elohim, sons of God, messengers, cherubim, seraphim, Watchers, giants, and a governing council.

The word elohim is grammatically plural in form, although Hebrew grammar can pair it with singular verbs when referring to Israel's primary deity. In other passages it refers to multiple divine beings. Its meaning depends upon context, and the surviving text preserves both singular authority and plural divine populations.

"Let us make"

Genesis 1:26 uses plural speech: "Let us make man in our image, after our likeness" in the Jewish Publication Society's 1917 translation. Verse 27 describes the act of creation with singular grammar. A heavenly-council address is one scholarly explanation of the plural; other interpretations emphasize deliberation or compositional history.[10][17]

In my interpretation, the wording preserves the memory that humans were created through the involvement of multiple beings. "Our image" refers to the humanoid resemblance shared by the Faceasian genetic contributors and the resulting human population.

The plural wording is present in the received text, but it does not make every occurrence of elohim grammatically plural. My biological reading of “image” is additional to the verse’s theological language, not an explicit reference to genetics.

Nations allotted among divine beings

Deuteronomy 32:8–9 preserves a significant variant. A Dead Sea Scroll witness has divine-son language, and the Septuagint tradition includes “angels of God,” while the Masoretic text reads “sons of Israel.” The Most High divides peoples; Jacob is then described as YHWH’s portion.[17][18] The relationship between these lines and older West Semitic divine-council traditions remains a subject of scholarship.

In my interpretation, this division preserves the assignment of human regions to different ruling beings or factions. Later monotheistic development subordinated the other members of the council beneath one supreme authority without removing every trace of the earlier structure.

Psalm 82 and judgment within the council

Psalm 82 depicts God standing within an assembly and judging other elohim. These beings receive criticism for governing unjustly and failing to protect the weak. The scene preserves multiple authorities and a dispute concerning their rule.

The psalm condemns unjust judgment and demands protection for the weak and needy. Divine beings are a prominent scholarly reading of the elohim addressed, while other interpretations understand human judges.[17] In my analysis, the scene preserves a distinction between possessing authority and exercising it rightly. Its criticism is directed toward the behavior of rulers, not merely their existence.

The Watchers and restricted knowledge

Genesis 6 describes the sons of God taking human women and producing unusual offspring associated with the Nephilim and ancient heroes. The passage is brief, but it preserves descent, reproductive contact, mixed offspring, and concern about the consequences.

1 Enoch expands this tradition through the Watchers. These beings descend, take human partners, produce giant offspring, and teach knowledge involving weapons, metalworking, cosmetics, celestial observation, and other specialized practices.

The resulting punishment centers on crossing controlled boundaries and transferring restricted knowledge. In my interpretation, the Watcher tradition preserves contact between nonhuman rulers or teachers and the human population. The offspring record biological compatibility or further experimentation, while the forbidden teachings represent knowledge given without the approval of the ruling order.

The Book of the Watchers is a pre-Christian Jewish apocalyptic work, not a report written at the time of the events it describes. Aramaic fragments survive among the Dead Sea Scrolls; the complete collection of 1 Enoch is preserved in Ge'ez.[19] It develops the shorter Genesis tradition and addresses evil, violence, illicit teaching, and the consequences of boundary crossing. I connect those themes with the tensions between nonhuman teachers and controlling authorities in my account.

Cherubim and seraphim

Cherubim guard sacred boundaries, support or accompany the divine throne, and possess composite forms. Seraphim appear as winged or serpentine beings associated with radiant presence and restricted access.

Later art often depicts cherubim as humanlike angels, but the older descriptions present more complex beings. Their forms may preserve biological differences, protective equipment, ceremonial clothing, or specialized guardian roles.

In my interpretation, some of these beings may correspond with nonhuman populations or genetically modified guardians remembered through Hebrew religious language.

Ezekiel's moving throne

Ezekiel describes a radiant moving structure arriving with wind, cloud, fire, sound, living beings, intersecting wheels, and a thronelike platform above them. The entire formation moves as an organized unit and produces an overwhelming sensory effect.

The beings and wheels operate together. The structure changes direction without ordinary turning, rises from the ground, and carries a seated figure above the platform.

In my interpretation, Ezekiel recorded an encounter with unfamiliar transportation technology using the vocabulary available to him. Wheels within wheels, gleaming surfaces, fire, noise, movement, and an elevated operator describe the observed components of a craft more closely than an ordinary wheeled chariot.

The text presents the event as a prophetic vision of divine glory, with living creatures, wheels, an expanse, and a throne. Those are its descriptive terms. I interpret the coordinated movement as a possible encounter with unfamiliar transport; I do not replace the passage with a supposed technical translation. Its religious and literary setting is part of the surviving account.

Transmission rather than a single rewriting

This does not require the claim that one complete polytheistic Bible was rewritten at a single moment. The transformation occurred gradually as stories were transmitted, political authority changed, one deity was elevated, and earlier beings were subordinated, renamed, or reclassified.

The specific manuscript variants and changing interpretations matter more than a claim that every later translation is corrupt. Older plural language survives alongside increasingly exclusive forms of worship. I see that tension as a possible trace of remembered authority becoming universalized, while the actual textual history contains many communities, periods, and choices rather than one editing event.

CHAPTER 16

Mesoamerica: Successive Creations and Restricted Sight

Maya and central Mexican traditions preserve distinct histories of making

Mesoamerican history survives through Maya inscriptions, codices, monumental art, central Mexican manuscripts, architecture, oral traditions, and colonial-era recordings of Indigenous accounts.

Colonial conquest destroyed many original books and institutions. The surviving record is therefore incomplete, but it continues to describe earlier populations, multiple creations, celestial rulers, underworld beings, culture-bringers, sacred calendars, and knowledge received from nonhuman sources.

The Popol Vuh

The Popol Vuh belongs specifically to K'iche' Maya tradition. Its surviving alphabetic form was written after the Spanish conquest, drawing on earlier narratives and remembered knowledge. It is a sacred and dynastic history, not a pre-Columbian tablet recording a laboratory procedure.[37]

The Maya Popol Vuh describes repeated attempts to create humans. Beings made from mud cannot function successfully. Wooden people live but lack the awareness and relationship expected by their makers. A later humanity is formed from maize and becomes capable of thought, speech, memory, and recognition.

The creators initially find the new humans too perceptive. Their sight extends too far, allowing them to understand more of reality than intended. Their vision is then clouded so that they can perceive only what is near.

In my interpretation, this is one of the clearest cultural memories of experimental variation followed by the deliberate restriction of human perception. The failed bodies preserve unsuccessful biological forms, while the clouding of sight corresponds with the reduction or deactivation of wider sensory and Source awareness.

Maize is central to K'iche' nourishment, identity, and relationships with the sacred. I do not translate it as ape DNA or treat it as a hidden biological label. My correspondence lies in the narrative sequence: makers try different forms, a successful humanity appears, and perception is then limited. The local meaning of maize remains essential even within that comparison.

Quetzalcoatl and knowledge

Quetzalcoatl, the Feathered Serpent, appears in different forms across Mesoamerican cultures. He is connected with wind, knowledge, priesthood, calendrics, creation, rulership, and the transmission of civilizing practices.

His combined avian and serpentine appearance may preserve a relationship between different lineages, specialized clothing, or a being whose form combined characteristics later represented through feathers and serpent anatomy.

Quetzalcoatl is remembered as an active teacher and culture-bringer rather than only as a natural force. In my interpretation, his stories preserve an embodied visitor or ruler associated with advanced knowledge.

Sky bands and celestial routes

Mesoamerican art places rulers and deities within sky bands, serpentlike structures, elevated platforms, caves, celestial roads, and openings between different regions. Canoes and litters also appear as forms of transportation.

In my interpretation, some serpent-shaped or celestial structures may represent craft whose movement and exterior forms were described through familiar animals. A long articulated vehicle moving through the sky could be remembered as a feathered or luminous serpent.

The surrounding forms belong to local systems of kingship, cosmic order, and ritual movement. I interpret some as possible memories of transport or equipment, while recognizing that the identification as technology is mine rather than the wording of an accompanying engineering inscription.

Pakal's sarcophagus lid

The sarcophagus lid of K'inich Janaab' Pakal at Palenque shows the ruler within an elaborate cosmological composition involving the world tree, celestial imagery, ancestral transition, and the entrance into another state of existence.

The inscriptions identify Pakal and connect the scene with Maya kingship and death. Its meaning within the tomb cannot be separated from that funerary setting.

At the same time, the arrangement resembles a person positioned within an operating structure, surrounded by controls, mechanical forms, and directed movement. In my interpretation, the imagery may combine Pakal's transition after death with inherited memories of beings operating advanced transportation systems.

The scene does not need to be either a religious image or a technological memory. Ancient people could express technology through cosmology, especially when movement between the sky, Earth, and the underworld had become part of sacred history.

Pakal was a historical Maya ruler. The lid's funerary context and inscriptions connect its imagery with kingship, ancestry, and death; they do not call him a spacecraft pilot.[38] My technological comparison concerns the arrangement I see in the scene, not a replacement identity for the king or a translation of the inscription.

CHAPTER 17

Greece and the Mediterranean: Families, Makers, and Machines

Succession, mixed ancestry, forbidden knowledge, and autonomous devices

Greek tradition preserves multiple generations of divine beings rather than one unchanging population. The primordial powers are followed by the Titans, and the Titans are overthrown by Zeus and the Olympians.

The conflict includes family relationships, imprisonment, alliances, specialized weapons, territorial division, and the replacement of one ruling generation by another. After victory, Zeus, Poseidon, and Hades divide authority over the sky, sea, and underworld.

In my interpretation, these succession wars preserve later human memories of conflicts among older nonhuman populations and ruling families. The assignment of territories resembles political administration after a transfer of power.

Hesiod's Theogony preserves the succession of divine generations, while other epic traditions supply additional accounts of territorial division and heroic ancestry. These are related but not identical literary sources. Their genealogies also served religious, poetic, and aristocratic purposes within Greek society.[39]

Mixed descendants

Greek gods possess bodies, reproduce, suffer injury, experience desire, disguise themselves, travel, and produce children with human partners. Heroes such as Heracles, Perseus, and others carry mixed divine and human ancestry.

The offspring inherit unusual strength, perception, abilities, or political importance. Their lives are shaped by both their human environment and their connection to a nonhuman parent.

In my interpretation, these accounts preserve memories of reproductive contact between humans and visiting rulers. Later storytelling amplified individual abilities, but the repeated structure of mixed ancestry remained central to heroic genealogies.

Prometheus and the gift of fire

Prometheus creates or assists humanity and gives humans fire against the wishes of Zeus. Fire represents physical flame, but it also becomes the foundation of cooking, protection, metallurgy, construction, and technological independence.

Zeus punishes Prometheus because the transfer changes the relationship between rulers and humans. A population with access to advanced knowledge becomes less dependent upon those controlling it.

This follows the same structure preserved in Mesopotamian and Hebrew traditions: one authority restricts knowledge, another teacher transfers it, and punishment follows. In my interpretation, Prometheus preserves the memory of a being who gave humans technical knowledge without the approval of the ruling faction.

Hephaestus and engineered attendants

Hephaestus is the divine metalworker, engineer, and builder of extraordinary devices. His creations include automated attendants, self-moving tripods, specialized armor, weapons, traps, and structures made for the gods.

The golden attendants described in the Iliad possess movement, intelligence, speech, and the ability to assist their creator. The self-moving tripods travel to and from divine gatherings without being pulled by animals or humans.

These accounts describe the concept of engineered artificial beings and autonomous machines long before modern robotics. In my interpretation, they preserve knowledge of advanced mechanical systems inherited from earlier technological civilizations.

Talos, the bronze guardian of Crete, is another example. He circles the island, identifies approaching threats, and attacks intruders. His body contains a vital internal channel closed by a plug or fastener. When it is opened, the substance sustaining him escapes and he becomes inactive.

The description resembles a constructed guardian whose operation depends upon a contained internal system. Talos may preserve the memory of an artificial protective machine rather than a biological giant.

The golden helpers and self-moving tripods appear in Book 18 of the Iliad; Talos is described in Book 4 of Apollonius' Argonautica.[40][41] Their literary settings establish that ancient writers imagined artificial agents and moving devices. My reading of them as remembered technology goes beyond what the existence of the poems alone demonstrates.

Vehicles, weapons, and realms

Greek deities travel in chariots, boats, clouds, and other extraordinary conveyances. Helios crosses the sky in a solar chariot. Athena, Apollo, and other gods arrive or depart through vehicles associated with light, speed, sound, and celestial movement.

Zeus's thunderbolt is a specialized weapon produced by the Cyclopes. Poseidon carries a trident capable of affecting land and sea, while other divine instruments produce concealment, transformation, protection, or destruction.

In my interpretation, these objects preserve advanced technologies remembered through the language of divine weapons. Their effects became attached to natural forces because lightning, earthquakes, storms, and fire provided the closest human comparisons.

Sea Beings and Subterranean Populations

Greek and Mediterranean traditions include intelligent sea gods, Tritons, Nereids, and other beings capable of living beneath the water while interacting with surface populations. Some possess humanlike upper bodies and aquatic lower forms.

These accounts may preserve distant memories of aquatic populations resembling the Fraedians or other beings equipped to survive underwater. Their family relationships, territories, vehicles, and political roles present them as organized persons rather than ordinary fish.

The underworld is likewise described as a governed region with entrances, boundaries, guardians, rivers, and rules controlling passage. Some of this imagery concerns death, but it may also preserve memories of restricted subterranean locations or enclosed environments governed by nonhuman authorities.

The divine families are not a literal chronological register of the Lyran war in my interpretation. I connect their recurring problems, inheritance, power, knowledge, and the treatment of descendants with the wider history proposed here. Their differences remain part of the Greek and Mediterranean traditions rather than details to be flattened into one genealogy.

CHAPTER 18

Older Images and Human Engineering

Jōmon figurines and the monumental world before written history

Dogū and remembered figures

Jōmon Japan produced clay figurines known as dogū over a long period. Their forms vary; the enlarged eyes of the “goggle-eyed” examples are one recognizable type, not a feature of every figure. Interpretations include ritual, healing, fertility, protection, and other social or spiritual purposes. No surviving Jōmon text identifies them as astronauts.[42]

The dogū figures of Jōmon Japan may similarly depict rulers, teachers, visitors, or specialized beings who appeared in that region. Their enlarged eyes, unusual head shapes, segmented bodies, fitted surfaces, and repeated nonhuman proportions may record actual anatomy, protective suits, technological equipment, or combinations of biological and engineered features. I interpret that resemblance as a possible memory of embodied visitors, although no surviving written explanation identifies the figures that way.

Many dogū were used in ritual settings and are associated conventionally with healing, fertility, protection, or spiritual practice. In my analysis, those rituals may have developed around memories of beings who served as healers, teachers, rulers, or intermediaries. If such beings possessed advanced biological knowledge, protective equipment, or transportation technology, later communities could have preserved their appearances and functions through figurines after the original practical context was lost.

The suit or equipment resemblance is the correspondence I draw, not a recovered operating description. Without a named textual identity, I do not place the figures within a specific lineage or make them participants in the human genetic project.

Göbekli Tepe

Figure 13. Animal reliefs on a pillar at Göbekli Tepe, c. 9600–8800 BCE. Photograph by Teomancimit, CC BY-SA 3.0, via Wikimedia Commons.

Göbekli Tepe in southeastern Türkiye preserves monumental enclosures begun in the tenth millennium BCE. T-shaped pillars carry animals and, in some cases, carved arms, hands, and belts. The anthropomorphic features distinguish those pillars from unworked stones, although the identities of the represented beings remain uncertain.[43]

The builders could organize quarrying, carving, transport, and collective construction before the urban states of Mesopotamia. The site demonstrates the depth of early human social and symbolic life. It does not require an outside builder to explain human competence, and it has no inscription naming the lineages used in this book.

In my interpretation, such an early symbolic world leaves room for memories older than surviving writing. Animals, humanlike forms, and monumental settings could carry stories through visual relationships rather than sentences. That possibility does not identify every carved animal as a biological hybrid. The image preserves an ancient community's work even where the story it told is no longer recoverable.

PART V - COMMUNICATION AND THE CONTINUING HUMAN STORY

CHAPTER 19

NSA Cryptology and Extraterrestrial Communication

A real paper about hypothetical messages-not a disclosure of contact

Communication with Extraterrestrial Intelligence is an unclassified paper by Lambros D. Callimahos, an NSA cryptologist. Its substance was presented during a panel at the 1965 IEEE Conference on Military Electronics in Washington, D.C. The panel included specialists in cryptology, linguistics, physics, astronomy, and dolphin communication. The NSA later released the paper among declassified technical-journal materials. Its presence on the NSA website establishes that a defense and cryptology community seriously discussed the problem of recognizing and decoding communication from a nonhuman civilization.

The paper is not a report that the NSA received a message. Callimahos constructs hypothetical signals to demonstrate how contact might proceed. He begins from the mid-twentieth-century scientific expectation that life-supporting worlds could be numerous, refers to Frank Drake and Project Ozma, and discusses listening near the neutral-hydrogen frequency of 1420.405752 megahertz. He considers radio, lasers, masers, and even speculative neutrino communication as media an advanced civilization might use.

Mathematics without a shared language

Callimahos argues that an unknown sender would first distinguish intention from cosmic noise. Repeated natural numbers, primes, geometric relations, and carefully factored binary message lengths could reveal design without requiring a shared spoken language. A rectangular raster could encode a picture once the receiver discovered the correct dimensions. The paper explains an imagined pictogram containing a star, planets, chemistry, bodies, water, a fish, height, and number-system clues. It then develops a longer invented code that teaches arithmetic and symbolic relationships step by step.

Callimahos describes the problem as inverse cryptography: instead of concealing meaning, a sender tries to make it recoverable by a recipient without a shared language or culture. A sequence can suggest deliberate structure before it reveals what the sender means. The hypothetical exercises in the paper illustrate that distinction; they are not intercepted alien messages.[44]

The risks of advanced civilization

The paper raises a question that parallels the Reptilian catastrophe without proving it. Callimahos observes that technologically advanced civilizations may destroy themselves. In a cited 1965 discussion, radio astronomer Iosif Shklovsky lists potential civilizational crises including thermonuclear self-destruction, genetic danger, information overload, extreme specialization, and the creation of artificial intelligent beings. These were human scientific concerns, not disclosures of Lyran history. Their relevance is conceptual: increasing the power to alter life and create dependent intelligence also increases the scale of possible failure.

I connect that concern with the broader theme of this book: technical power can outgrow the judgment used to direct it. Genetic intervention and artificial intelligence create possibilities for preservation as well as harm. The NSA paper discusses human scientific speculation about those possibilities; it provides no confirmation of the Lyran experiments or the Reptilian war.

Its contribution is a historical one. Specialists treated communication with an unknown intelligence as a serious technical problem involving signal, structure, biology, and interpretation. The presence of the paper in a government archive does not change a hypothetical exercise into evidence of an exchange. It extends the discussion from ancient records to modern attempts to imagine how meaning might cross the distance between different kinds of mind.

Conclusion - From Source to Humanity

The creation story I have presented moves from a question about consciousness to a question about responsibility. I interpret Source as the common origin of existence, and biological lives as distinct perspectives within it. The Lyran and Faceasian accounts describe what happens when the capacity to make becomes entangled with the authority to command. The historical chapters place that interpretation beside traditions of divine families, workers, teachers, craft, and restricted knowledge.

These records do not establish the complete off-world lineage. They preserve the words, images, beliefs, and achievements of the human cultures that transmitted them. Their value does not disappear when an interpretation remains uncertain. They reveal how deeply people have considered the difference between being created, being governed, and being free.

The spiritual meaning does not depend upon winning an argument about ancient visitors. If consciousness is rooted in Source, no creator species owns the beings that arise through it. A body may be designed, but awareness cannot be reduced to a patent. The deepest error of the lineage is not biological experimentation by itself; it is the conversion of relationship into property. The deepest human task is not to exchange worship of an invisible god for worship of advanced ancestors. It is to recover direct responsibility for what we create and how we treat every form through which consciousness looks back at us.

Humanity stands at the threshold once occupied by its proposed makers. We alter genes, build artificial minds, design dependent systems, and reshape ecosystems. We can repeat the old patterns created for service, restrict knowledge, and punish independence or transform it. To become worthy participants in creation means allowing what we bring forth to possess value beyond its usefulness to us. It means joining knowledge with restraint and power with relationship.

For me, remembering Source is not obedience to another powerful ancestor. It is recognition that consciousness has value beyond usefulness. Humanity continues the story whenever it creates knowledge, changes a living system, or decides how another being should be treated. The future does not have to repeat the hierarchy through which I interpret the past.

Glossary

Source

The name used in this cosmology for undivided conscious potential prior to separate observers and durable matter. It is a metaphysical concept, not a synonym for a measured physical field.

Plasma

Ionized matter containing free charged particles and displaying collective electromagnetic behavior. Plasma forms stars, the solar wind, lightning, and many laboratory discharges. Its responsiveness is established; consciousness in plasma is not.

Resonance

Selective response of a system to particular modes or frequencies. I also use resonance as an analogy for coherent participation; that broader use is not itself a frequency measurement.

Lineage

A proposed chain of biological and conscious inheritance. Some lineage names belong to the account I received rather than ancient texts.

Lyran

The first advanced humanoid civilization in my interpretation, established in an unidentified planetary system in the direction now called Lyra.

Faceasian

The received name for a power-holding maker lineage correlated here with the Mesopotamian ruling-god complex. The correlation is not a translation.

Anunna or Anunnaki

A changing group designation within the Mesopotamian pantheon, sometimes especially connected with underworld gods in later contexts. It does not linguistically mean extraterrestrials who came to Earth.

Igigi

A Mesopotamian group of gods whose exact relationship to the Anunna varies by text. In Atrahasis-related interpretation they are especially relevant to divine labor.

Elohim

A Hebrew form morphologically plural but capable of singular grammar when referring to the God of Israel; it can also refer to other divine beings. Meaning depends upon syntax and context.

Divine council

A sacred assembly in which a high god presides among other divine beings. Councils occur across the ancient Near East and may reflect theology, kingship, political memory, or several layers at once.

Vimana

A Sanskrit term with meanings that vary by period and context, including palace, elevated structure, celestial conveyance, or temple superstructure. The word alone does not identify an aircraft.

Apkallu

Mesopotamian sages associated with ancient knowledge and protection, represented in several forms, including human, fish-cloaked, winged, or eagle-headed figures depending on context.

Hybrid

In my reconstruction, a population integrating inheritance from different biological lines. In ancient art, a composite body may instead be symbolic, ritual, transformational, or protective.

Reconstruction

My proposed synthesis of received lineage history with comparative evidence. It is explicitly distinguished from documented fact.

Sources and Suggested Reading

Bracketed numbers identify the references below. Ancient texts are generally summarized in the chapters; direct quotations identify a translation. Photo credits remain with the figures. The sources document their own scientific, historical, or literary subjects and do not independently verify the proposed off-world lineage.

[1] NASA Science. 'Our Sun: Facts.' https://science.nasa.gov/sun/facts/

[2] U.S. Department of Energy/NNSA. Stewardship Science Academic Programs Annual (2014), discussion of self-organized electromagnetic structures and filaments. https://www.energy.gov/sites/prod/files/migrated/nnsa/2017/11/f44/2014-02-28%20SSAP%20Annual_February%202014_print.pdf

[3] ORACC, Ancient Mesopotamian Gods and Goddesses. 'Igigi/Igigu.' https://oracc.museum.upenn.edu/amgg/listofdeities/igigi/index.html

[4] Electronic Text Corpus of Sumerian Literature. 'Enki and Ninmah' (ETCSL 1.1.2). https://etcsl.orinst.ox.ac.uk/cgi-bin/etcsl.cgi?text=t.1.1.2

[5] ORACC, Ancient Mesopotamian Gods and Goddesses. 'Anunna (Anunnaki).' https://oracc.museum.upenn.edu/amgg/listofdeities/anunna/index.html

[6] The Metropolitan Museum of Art. 'Divine Egypt: Inside the Exhibition.' https://www.metmuseum.org/exhibitions/divine-egypt/inside-the-exhibition

[7] Madain Project. 'Dendera Light' (photographic/context summary). https://madainproject.com/dendera_light

[8] Peter J. Brand, Ramesses II, Egypt's Ultimate Pharaoh (Lockwood Press, 2023), pp. 82–84, 108; see also American Research Center in Egypt, Temple of Seti I at Abydos, https://arce.org/temple-seti-i-abydos/

[9] Kisari Mohan Ganguli, trans. The Mahabharata of Krishna-Dwaipayana Vyasa (1883–1896), public-domain text index. https://sacred-texts.com/hin/maha/index.htm

[10] The Holy Scriptures According to the Masoretic Text, Jewish Publication Society (1917), Genesis 1:26–27. Hebrew text and translations: https://www.sefaria.org/Genesis.1.26

[11] TheTorah.com. 'The Seraphim,' including Hebrew divine-council passages in Psalms and Job. https://www.thetorah.com/article/the-seraphim

[12] ORACC, Ancient Mesopotamian Gods and Goddesses, deity index. https://oracc.museum.upenn.edu/amgg/listofdeities/index.html

[13] British Museum. Wall panel depicting a protective spirit/apkallu, object W_1849-0502-2. https://www.britishmuseum.org/collection/object/W_1849-0502-2

[14] Princeton University Art Museum. Relief of a winged genie/apkallu. https://artmuseum.princeton.edu/art/collections/objects/34202

[15] Stephanie Dalley, trans. Myths from Mesopotamia: Creation, the Flood, Gilgamesh, and Others, rev. ed. (Oxford University Press, 2000).

[16] Benjamin R. Foster, Before the Muses: An Anthology of Akkadian Literature, 3rd ed. (CDL Press, 2005).

[17] Mark S. Smith, The Origins of Biblical Monotheism (Oxford University Press, 2001).

[18] Emanuel Tov, Textual Criticism of the Hebrew Bible, 4th ed. (Fortress Press, 2022).

[19] George W. E. Nickelsburg and James C. VanderKam, 1 Enoch: The Hermeneia Translation (Fortress Press, 2012).

[20] John H. Walton, Ancient Near Eastern Thought and the Old Testament, 2nd ed. (Baker Academic, 2018).

[21] Jan Assmann, The Search for God in Ancient Egypt (Cornell University Press, 2001).

[22] Erik Hornung, Conceptions of God in Ancient Egypt: The One and the Many (Cornell University Press, 1982).

[23] Valmiki. The Ramayan of Valmiki, translated by Ralph T. H. Griffith (1870–1874), especially the Yuddha Kanda return journey. Public-domain translation: https://sacred-texts.com/hin/rama/

[24] Stella Kramrisch, The Hindu Temple (University of Calcutta, 1946), for the architectural meanings of vimana.

[25] Chimpanzee Sequencing and Analysis Consortium. “Initial sequence of the chimpanzee genome and comparison with the human genome.” Nature 437 (2005): 69–87. https://doi.org/10.1038/nature04072

[26] NASA Goddard Space Flight Center, Imagine the Universe!, 'Sheets and Voids,' explaining that galaxies form sheets and filaments around voids thought to fill about 95 percent of space. https://imagine.gsfc.nasa.gov/features/cosmic/sheets_voids_info.html

[27] U.S. Department of Energy, Hydrogen Properties, explaining that atoms consist largely of open space relative to the scale of their nuclei and electron distributions. https://www.energy.gov/documents/fcm01r0pdf

[28] CERN, 'What's so special about the Higgs boson?', explaining the quantum-field description in which fundamental fields fill the universe and particles are excitations of those fields. https://home.cern/science/physics/higgs-boson/what/

[29] Campante, T. L., et al. “An Ancient Extrasolar System with Five Sub-Earth-Size Planets.” The Astrophysical Journal 799 (2015): 170. https://doi.org/10.1088/0004-637X/799/2/170

[30] Gracheva, E. O., et al. “Ganglion-specific splicing of TRPV1 underlies infrared sensation in vampire bats.” Nature 476 (2011): 88–91. https://doi.org/10.1038/nature10245

[31] Ijdo, J. W., et al. “Origin of human chromosome 2: an ancestral telomere-telomere fusion.” Proceedings of the National Academy of Sciences 88 (1991): 9051–9055. https://doi.org/10.1073/pnas.88.20.9051

[32] Gracheva, E. O., et al. “Molecular basis of infrared detection by snakes.” Nature 464 (2010): 1006–1011. https://doi.org/10.1038/nature08943

[33] Kawashima, Y., et al. “Mechanotransduction in mouse inner ear hair cells requires transmembrane channel-like genes.” Journal of Clinical Investigation 121 (2011): 4796–4809. https://doi.org/10.1172/JCI60405

[34] British Museum. Adda Seal, collection number 89115. See also World History Encyclopedia, “Enki,” for the reproduced detail used in this manuscript: https://www.worldhistory.org/Enki/

[35] Verbrugghe, Gerald P., and John M. Wickersham. Berossos and Manetho, Introduced and Translated: Native Traditions in Ancient Mesopotamia and Egypt. University of Michigan Press, 1996.

[36] Collon, Dominique. The Queen of the Night. British Museum Press, 2005.

[37] Christenson, Allen J., trans. Popol Vuh: Sacred Book of the Quiché Maya People. University of Oklahoma Press, 2007.

[38] Schele, Linda, and Peter Mathews. The Code of Kings: The Language of Seven Sacred Maya Temples and Tombs. Scribner, 1998. Palenque and the Temple of the Inscriptions.

[39] Hesiod. Theogony and Works and Days, translated by M. L. West. Oxford University Press, 1988.

[40] Homer. The Iliad, Book 18, translated by Richmond Lattimore. University of Chicago Press, 1951.

[41] Apollonius of Rhodes. Argonautica, Book 4, translated by R. C. Seaton. Loeb Classical Library, 1912.

[42] Habu, Junko. Ancient Jomon of Japan. Cambridge University Press, 2004.

[43] Schmidt, Klaus. “Göbekli Tepe-the Stone Age Sanctuaries. New results of ongoing excavations with a special focus on sculptures and high reliefs.” Documenta Praehistorica 37 (2010): 239–256.

[44] Callimahos, Lambros D. “Communication with Extraterrestrial Intelligence.” Paper based on a panel at the IEEE Conference on Military Electronics, September 23, 1965. NSA technical-journal archive: https://www.nsa.gov/portals/75/documents/news-features/declassified-documents/tech-journals/communications-extraterrestrial-intelligence.pdf